THE INT[illegible]
of
NATURE and the PSYCHE

C. G. JUNG

SYNCHRONICITY:
AN ACAUSAL CONNECTING
PRINCIPLE

W. PAULI

THE INFLUENCE OF ARCHETYPAL
IDEAS ON THE SCIENTIFIC
THEORIES OF KEPLER

The Interpretation of Nature and the Psyche: Synchronicity- An Acausal Connecting Principle / The Influence of Archetypal Ideas on the Scientific Theories of Kepler

by Carl Gustav Jung (1875-1961)

and Wolfgang Ernst Pauli (1900 – 1958)

First published in 1952 in German as Naturerklärung und Psyche. C. G. Jung: Synchronizität als ein Prinzip akausaler Zusammenhänge. W. Pauli: Der Einfluss archetypischer Vorstellungen auf die Bildung naturwissenschaftlicher Theorien bei Kepler

This Printing in February, 2012 by Ishi Press in New York and Tokyo with a new introduction by Marvin Jay Greenberg

ISBN 4-87187-713-2

978-4-87187-713-8

Ishi Press International

1664 Davidson Avenue, Suite 1B

Bronx NY 10453-7877

USA

1-917-507-7226

Printed in the United States of America

Foreword to the Ishi Press Edition

By Marvin Jay Greenberg, Ph.D.

Scientific and Personal Information about Wolfgang Pauli

Wolfgang Pauli (1900–1958) was a leading twentieth century quantum physicist. He was awarded a Nobel Prize for his *Exclusion Principle.*

Wolfgang Pauli

Pauli came up with the idea that an electron had four quantum numbers determining its energy state, not three as was previously believed, and that the new fourth number could only take on the values plus or minus one half. Later that fourth number was shown to represent the *spin* of the electron. His *Exclusion Principle* stated that two electrons in an atom could not have the same four quantum numbers. It accounts for the periodic table of elements.

From the Nobel Prize committee statement about that:

> "Pauli showed himself that the electron configuration is made fully intelligible by the *Exclusion Principle*, which is therefore essential for the elucidation of the characteristic physical and chemical properties of different elements. Among those important phenomena for the explanation of which the Pauli principle is indispensable, we mention the electric conductivity of metals and the magnetic properties of matter."

Particles have been classified into two types: *bosons*, whose wave function is symmetric under exchange, and *fermions*, whose wave function is anti-symmetric under exchange. All particles have either integer spin or half integer spin in units of the reduced Planck constant. The *Spin Statistics Theorem* states that the particles with integer spin are the bosons and the particles with half integer spin are the fermions. Markus Fierz (1912–2006), Pauli's doctoral student, first formulated this theorem and Pauli re-derived it in a more systematic way. It is only the fermions that are subject to Pauli's *Exclusion Principle*. Fermions obey Fermi-Dirac statistics while bosons are subject to Bose-Einstein statistics.

In pondering the loss of energy in beta-decay, knowing that conservation of energy was built into the quantum electrodynamics equations he had published with Werner Heisenberg (1901–1976), Pauli conjectured in 1930 that the nucleus undergoing beta-decay emitted a new particle of zero mass and zero charge - later named by Enrico Fermi (1901–1954) the *neutrino* - whose existence and properties would restore conservation of energy. It wasn't until 26 years later that physicists were able to experimentally verify the existence of the neutrinos Pauli envisioned.

T. D. Lee and C. N. Yang had conjectured that parity might not be conserved in the case of weak interactions; C-S. Wu headed a group that experimentally verified such overthrow of parity. She might have been the "concretization" (in the terminology of Rolf R. Loehrich) of the Chinese woman appearing in Pauli's dreams about mirrors and their reflections. Pauli was astonished at this discovery.

Pauli considered instead a fuller symmetry, *CPT symmetry* (charge conjugation, parity, time reversal). His theorem asserts roughly that a mirror universe – in which all matter is replaced by antimatter, all positions are their own reflections and time runs backward – would be indistinguishable from our own universe. To date no violations of CPT symmetry have been found experimentally.

Pauli's many stimulating, critical discussions with Heisenberg were very important in the latter's discoveries. They also collaborated on a theory of quantum electrodynamics that was later supplanted by Richard Feynman's diagrams (sarcastically called "sentimental paintings" by Pauli).

Pauli became known as "the conscience of physics" because of his insistence on rigor and his severe criticism of any theory or lecture that was not up to his high standards. Top level physicists like Heisenberg, Bohr and Einstein valued his criticism, forcing them to strengthen their theories or else eventually abandon them; lesser physicists found his sarcastic comments offensive. His most famous such insulting criticism is "Not only is it not right, it is not even wrong!" Another: "The fact that the author thinks slowly is not serious, but the fact that he publishes faster than he thinks is inexcusable." The physicist Paul Ehrenfest (1880–1933) called Pauli "der Geissel Gottes" (the scourge of God) because of the way Pauli criticized a presentation by Ehrenfest and was so critical generally. Heisenberg claimed never to have published without first having Pauli check his work.

When Pauli's assistant Victor Weisskopf (1908–2002) published an article that contained an error, Pauli stopped speaking to him and refused to see him for a while. Later, Weisskopf wrote:

> "All of Pauli's disciples developed deep personal attachment to him, not only because of the many insights he gave us, but because of his fundamentally endearing human qualities. It is true that sometimes he was a little hard to take, but all of us felt that he helped us to see our weaknesses.... Pauli's occasional and highly publicized roughness was an expression of his dislike of half-truths and sloppy thinking, but it was never meant to be directed against any person. Pauli was an excessively honest man: he had almost a childlike honesty. What he said were always his true thoughts, directly expressed."

This characteristic of Pauli's made it very surprising that he accepted Carl Jung's psychological/philosophical theories, for which he did not demand the same standard of rigor. But Pauli was critical of many of

Jung's ideas in private letters to him.

Pauli was Jewish and escaped from potential Nazi influence in Switzerland to spend several years at the Institute for Advanced Study in Princeton with his friend Albert Einstein. Eventually the Institute offered Pauli a permanent appointment, like Einstein's. He declined the offer in favor of returning to the ETH in Switzerland after the war.

Pauli went through a period starting at age 23 of self-destructive personal behavior. While at Hamburg University working hard on research in physics by day, Pauli secretly roamed the bars and brothels of the red light district — synchronistically named *Sankt Pauli* — at night. He had a short affair there with a woman who turned out to be a heroin addict. He drank too much and got into brawls. He wrote years later: "I tended toward being a criminal, a thug …" After he moved to a position in Zurich, he still returned frequently to Sankt Pauli.

Pauli also explored the demi-monde of Berlin. It was there that he met and fell for a cabaret dancer who was a friend of his sister. Pauli married her despite her letting it be known that she was in love with someone else. The marriage was a disaster and they divorced in less than a year. She soon remarried to a chemist, which was a further insult to Pauli, who considered chemistry an inferior science. Remarkably Pauli continued excellent work in physics despite his miserable personal life.

At age 31, on a summer lecture tour around the U.S. during the prohibition era, Pauli managed to obtain smuggled liquor from nearby Canada while he was in Ann Arbor. Then at a dinner party he was so drunk he fell down a flight of stairs and broke his shoulder. His injured arm was put in a device with a metal rod raising his arm; he joked that it was the only time in his life when he saluted "Heil Hitler."

After that U.S. tour, back in Switzerland, he went on a drinking binge, resuming his life of barroom brawls and whoremongering. He also got into bitter quarrels with colleagues at the ETH, leading to a reprimand from the administration.

Finally Pauli accepted the advice of his hated father (who had left his mother for a younger woman, with his mother subsequently committing suicide): Pauli sought help for his emotional condition from the famous psychoanalyst Carl Jung (1875–1961).

Therapy with Jung

In Jung's theory of personality types, a person was either an extrovert or an introvert. There were four main personality functions in opposing pairs: thinking vs. feeling and intuiting vs. sensing. (Rolf R. Loehrich, who studied with Jung, later added a fifth main function: integrating.) Jung's description of the introverted-thinking type fit Pauli at that time to a T:

> "His judgment appears cold, obstinate, arbitrary and inconsiderate; only with difficulty can he persuade himself to admit that what is clear to him may not be clear to everyone; if he happens to fall among people who cannot understand him, he proceeds to gather further proof of the unfathomable stupidity of man; he may develop into a misanthropic bachelor with a childlike heart; he appears prickly, inaccessible, haughty; has a vague dread of the other sex."

Four years later, Jung revealed this description of Pauli on first meeting him but without naming him:

> "He is a highly educated person with an extraordinary development of the intellect, which was, of course, the origin of his trouble – he was just too one-sidedly intellectual and scientific. He has a most remarkable mind and is famous for it. He is no ordinary person. The reason he consulted me was that he had completely disintegrated on account of this very one-sidedness. It unfortunately happens that such intellectual people pay no attention to their feeling life and so they lose contact with the world that feels, living in a world that thinks, a world merely of thoughts. So in all his relations to others and to himself he had lost himself entirely. Finally he took to drink and such nonsense and grew afraid of himself, could not understand how it happened, lost his adaptation, and was always getting into trouble. This is the reason he decided to consult me."

Later, in the preface to *Psyche and Symbol*, Jung added:

> "When the hard-boiled rationalist mentioned above came to consult me for the first time, he was in such a state of panic that not only he but I myself felt the wind blowing over our way from the lunatic asylum!"

Pauli told Jung that his dreams – full of seventeenth century style symbols and thinking – were driving him to distraction. Jung realized that not only was this young man in need of help but he was "chock full of archaic material." To keep that material free of direct influence from Jung himself, and also because Pauli needed connection to a supportive female, Jung refused initially to treat Pauli, referring him instead to his student Erna Rosenbaum.

Jung's intuition was that Pauli could never let down his defenses in front of a man, but with a woman he would be much freer to express himself. That was exactly what happened, as Pauli even rolled around on the floor while emotionally telling Rosenbaum his life stories. After five months therapy with Erna Rosenbaum, Pauli worked on his dreams alone for three months.

Finally, Jung agreed to work with Pauli the following year. By the time they concluded their sessions a year after that, Pauli had submitted 400 dream reports, including drawings as well as words, to Jung. Jung later wrote about those reports: "They contain the most marvelous series of archetypal images." Jung published a commentary on 72 of Pauli's 400 dream reports recorded before he was analyzed by Jung. They illustrate the process by which a patient is confronted with previously repressed parts of his nature that needed to be integrated to build a balanced personality.

But at Pauli's insistence, Jung never publicly revealed the identity of that dreamer. Pauli's identity as that dreamer and patient was only publicly revealed fifty years later, by Jung's secretary Aniela Jaffe.

In his many dreams, Pauli was sometimes confronted with an inner voice. Jung: "[Pauli] had the great advantage of being neurotic, so whenever he tried to be disloyal to his dreams or to deny the voice, the neurotic condition instantly came back."

Jung's method, in part, was to use dream reports and *active imagination* to enable the patient to meet his *shadow* – his dark side – and to separate it from his *anima* - his female aspect, the mediator between the conscious and the unconscious. Pauli's anima first appeared in a dream as a veiled woman, then later as an unknown woman, standing on a globe, worshipping the sun. The intense struggle between opposites in dreams and visions could resolve in the integration (cf. Loehrich) of his light and dark, good and evil sides, developing a more balanced personality. Jung called the completion of this process *individuation*, an ideal state in which all the psychological functions are fully conscious and form an integrated whole. Jung also showed Pauli that often images in his dreams could be found in old alchemical, religious or mythological texts (that Jung had collected) – they were images from *the collective unconscious*, postulated by Jung. For example, Pauli dreamt of a serpent biting its own tail; that was the mythological creature Uroboros, who symbolizes the eternal process of birth and rebirth. Jung interpreted this dream as first evidence of a transformation in Pauli. Rites of transformation involving serpents are ancient archetypes, appearing for example in Gnostic ceremonies of healing.

Pauli's dream report entitled "The House of Gathering" ended with "As I leave the house, I see a burning mountain and I feel 'The fire that cannot be put out is a holy fire.' According to Jung, Pauli's feeling that the unquenchable fire was 'holy' was a *sine qua non* of his cure. Then Pauli reported a vision – not a dream – of the "World Clock" that left him with a feeling of 'sublime harmony.' Jung wrote "This vision was a turning point in his psychological development. It is what one would call – in the language of religion – a conversion." Pauli described his transformation in a letter to his friend Kronig:

> "I became acquainted with psychic things, which earlier I did not know, which I summarize under the name 'independent activity of the soul.' That there are things here that are spontaneous products for growth, that can be designated as symbols of an objective psyche, things that cannot and should not be explicable from material causes, holds true for me without doubt."

After two years, Pauli ended his therapeutic work with Jung. Pauli did continue to send his dream reports to Jung. Their subsequent interactions were as collaborators. Pauli wrote to Jung:

> "The specific threat to my life has been that in the first half of life I swung from one extreme to the other … I was a cold and cynical devil to other people, a fanatical atheist and intellectual 'intriguer.' The opposition to that was, on the one hand, a tendency toward being a criminal, a thug (which could have degenerated into me becoming a murderer) and, on the other hand, being detached from the world – a totally unintellectual hermit with outbursts of ecstasy and visions."

Thanks to his work with Jung, Pauli in later years was somewhat calmer, staying out of trouble. He was able to make a new marriage to Franca Bertram and sustain it comfortably for the rest of his life. Pauli wrote to Jung that Franca "fell in love with my shadow side because it secretly made a great impression on her." (Strangely, when the newlyweds accidentally encountered Jung at a dinner party, Jung totally ignored Franca. She was negative toward Jung and insisted that Pauli end his therapy. After Pauli died, Franca destroyed all his dream reports she could find. She did not want the letters with Jung published. However, over two decades worth of letters between Pauli and Jung are now public.)

In his final letter to Jung before departing Switzerland for the U.S. to escape from a possible invasion by the Nazis, Pauli compared the psychological process of individuation with "the embryonic development from lower animals with single-blood vessels up to the formation of the heart." Pauli expressed thereby a disagreement with Darwin, arguing that the process of evolution was goal-directed toward completeness, not just random natural selection.

Collaboration with Jung

Pauli became intensely interested in developing a new perspective that brought together physics with Jung's analytical psychology yet transcended them both. His dreams were full of mysterious symbols, some related to physics. He wrote to Jung "It will become a matter of life and death for me to understand more about the objective meaning of these symbols than I do at the moment."

Pauli argued that we are innately born with certain organizing principles in our minds, certain archetypes that function as a connector between sense perceptions and concepts. He believed that his dreams provided evidence of that. Examples: He often dreamt about the fine structure of spectral lines, which reminded him of the lines drawn in the *I Ching*, which Pauli consulted (and which Jung took to be founded on the *synchronicity principle*). Pauli dreamt of doublets, which he consciously related to Bohr's complementarity theory of wave vs. particle as well as to other dualities.

Pauli found similarity between the 'stranger' in his dreams and the wizard Merlin from the legend of the Holy Grail. He wrote:

> "When rational methods in science reach a dead end, a new lease on life is given to those contents that were pushed out of time consciousness in the 17th century and sank into the unconscious. The 'stranger' happily uses the terminology of modern science (radioactivity, spin) and mathematics (prime numbers) but does so in an unconventional manner. Inasmuch as he ultimately wishes to be understood but has yet to find his place in our contemporary culture, he is, like Merlin, in need of redemption.... I want to recognize Merlin, talk to him again, bring his redemption a little nearer. That, I believe, is the myth of my life."

Pauli's essay about Kepler in this book synthesizes his thinking on this subject over a period of 25 years. His focus is particularly on the irrational aspect of the process of scientific creativity. He argued that this process involved connecting "inner images pre-existing in the human psyche" with external objects, and that "pure logic is incapable of constructing such a link." He claimed that the links between sense perceptions and concepts are *archetypes* – a word used by Kepler as well as Jung.

Pauli wrote:

> "It is only a *narrow* passage of truth (no matter whether scientific or other truth) that passes between the Scylla of a blue fog of mysticism and the Charybdis of a sterile rationalism. This will always be full of pitfalls and one can fall down on both sides."

Pauli considered that Bohr's concept of *complementarity* (referring to wave-particle duality) was an important step forward to achieve a reconciliation between those opposites. He wrote:

> "It would be most satisfactory of all if physics and psyche could be seen as complementary aspects of the same reality."

He claimed that *true philosophy begins with a paradox*. Complementarity accepts the paradoxical quality of nature. He claimed that the duality between the conscious and the unconscious was analogous. Neither matter nor psyche can ultimately be understood rationally. For Pauli, the solution to the psycho-physical problem lay in realizing that the scientist and the science constituted a whole.

In Jung's response to Pauli, he said "The uniting of opposites is not only an intellectual affair ... only out of one's totality can a person create a model of wholeness." He reminded Pauli that the archetype of wholeness was not yet within his reach.

Pauli's godfather Ernst Mach (1838–1916) regarded *metaphysics* – that which is not perceivable by the five senses – as the cause of all the evil on Earth. Pauli accepted that attitude until he was influenced by Jung.

For Pauli, the moral dilemma physics faced in the aftermath of the atom bomb was symptomatic of a concern that physics (and science in general) needed to expand its domain beyond the realm of rationally understood phenomena, needed to merge with the psychology of the unconscious, because its rationalistic perspective had fostered a dangerous "will to power" (cf. Francis Bacon's sixteenth century slogan 'knowledge is power'). He wrote "... the anxious question presents itself to us whether this power, our Western power over nature, is evil."

Two years after the publication in German of the Jung-Pauli collaborative book, Pauli wrote to Fierz:

> "Many physicists and historians have of course advised me to break the connection between my Kepler essay and C. G. Jung ... but *this dream symbolism makes an impact! The book itself is a fateful 'synchronicity'* and must *remain* one. I am sure that defiance would have unhappy consequences as far as I am concerned. Dixi et salvavi animam meam!" [I spoke and thus saved my soul.]

Jung's theory treated synchronistic events as *acausal*; he interpreted Heisenberg's uncertainty principle as indicating that there could be other connections of events besides the causal ones. E.g., how else explain instances of telepathy?

Jung underwent tough criticism of his theory from Pauli. Example: What Jung wrote about radioactivity was total nonsense to a physicist. Pauli differed from Jung's understanding of synchronicity insofar as Pauli firmly separated it from processes in physics. Jung wrote to Pauli, after they had met to discuss Jung's essay, that he had felt overwhelmed by Pauli's mathematically expressed thoughts about synchronicity. He said mathematics was for him like a "bottomless fog." Pauli told an associate of Jung that "In general I am now somewhat tired of the lack of education in mathematics and natural science in Jung's whole circle…."

Pauli cited *mathematics* as an example of application of ideas from the unconscious outside of psychology. Kepler considered geometry to be "the archetype of the beauty of the universe." Pauli also cited the work of his teacher Arnold Sommerfeld (1868–1951), who searched for simple empirical laws at the atomic level governed by whole numbers. Pauli said the general idea of the archetype should include primitive mathematical intuition. In his 1954 *Dialectica* article "Ideas of the Unconscious from the Standpoint of Natural Science and Epistemology," Pauli wrote: "The further development of ideas about the unconscious will not take place just within the narrow framework of their therapeutic applications but will be determined by their assimilation to the mainstream of natural science as applied to vital phenomena."

Pauli died of pancreatic cancer in December 1958 in hospital room 137, the (approximate) reciprocal of the fine structure constant α. Frau Pauli reported that his last words were "Now I would like still to speak only to one person: Jung."

Pauli's Kepler Essay in this book

Pauli sent Jung a dream report in which a blond man, a stranger, says he is seeking a neutral language that goes beyond terms such as 'physical' or 'psychic'; the dream was also about Kepler. Pauli wept in that dream. He interpreted it as modern scientists losing touch with their feeling function. That particular dream started Pauli on his study of Kepler. He sought to analyze the period of transition from mysticism and alchemy to the new rational scientific thinking.

Robert Fludd, a prominent alchemist and Rosicrucian, composed a violent polemic against Kepler's chief work *Harmonices Mundi* [Harmony of the World]. He said "it is for the vulgar mathematicians to concern themselves with quantitative shadows." For Kepler, only that which is capable of quantitative, mathematical proof belongs to objective science; the rest is personal.

Pauli distinguished two types of mind throughout history: "one type considering the quantitative relations of the *parts* to be essential, the other the qualitative indivisibility of the *whole*." Pauli called the first type "thinking-sensing," the second type "feeling -intuitive" according to Jung's classification of psychological types. Kepler and Fludd exemplify those two types.

Pauli supported Fludd's emphasis on the number four, which he called "the number of wholeness," in contrast to Kepler's inspiration from the Christian Trinity. (Recall Pauli's discovery of the fourth quantum number.) Appendix 2 reproduces Fludd's essay on the Quaternary. The alchemists focused on the four elements (earth, water, air, fire). The Chinese had five elements (metal and wood, no air), just as Loehrich had five primary archetypes he called "Rulers."

Fludd saw the world as more than simply a mechanical system. Fludd felt that scientific thinking was a threat to the "world soul." Pauli, observing the outcome three centuries later, felt that Fludd's fear was justified. Quantum physics showed that matter at its very root cannot be rationally understood. He wrote to Fierz in 1953 "I am not only Kepler but also Fludd."

On pages 208–212, concluding the main body of his essay, Pauli presented at length his own complex viewpoint, including:

> "The employability of old alchemical ideas in the psychology of Jung points to a deeper unity of physical and psychological occurrences. To us, unlike Kepler and Fludd, the only acceptable point of view appears to be the one that recognizes *both* sides of reality – the quantitative and the qualitative, the physical and the psychical – as compatible with each other, and can embrace them simultaneously."

From Pauli's letter to his sister Hertha in 1957:

> "I do not believe in the possible future of mysticism in the old form. However, I do believe that the natural sciences will out of themselves bring forth a counter pole in their adherents, which connects to the old mystic elements. On this subject

I have tried to find general formulations in my paper on Kepler, as also in a lecture on *Science and Western Thought*."

The "Pauli Effect"

Pauli's presence sometimes mysteriously resulted in things breaking or otherwise going wrong. This bizarre regularity was named *the Pauli effect*. A few of many examples: Pauli was visiting the Bergedorf observatory when a terrible accident befell the great refractor telescope. In a laboratory at Göttingen University, an apparatus for the study of atoms suddenly collapsed without apparent cause; the time this occurred was when the train carrying Pauli to Copenhagen from Zurich stopped for a few minutes at the Göttingen station. On another occasion, Pauli was on a train when the rear cars decoupled and were left behind while he proceeded to his destination in one of the front cars. When Pauli attended the opening ceremony for the C. J. Jung Institute in Zurich, a vase overturned, spilling water. Pauli was riding in a tram with three other physicists heading to Bellevue Square in Zurich; just as they reached the square, two other street cars collided right in front of theirs.

Whenever these catastrophes occurred, Pauli himself was not harmed, which may be considered a second kind of Pauli "exclusion principle." Pauli's former student, Robert Oppenheimer, rejected Pauli's offer to participate in the Manhattan Project to build an atomic bomb, possibly because of the Pauli effect.

Jung's Essay on Synchronicity

Here I will simply quote from the book by Joseph Cambray listed in the References:

Synchronicity as "a meaningful coincidence" and "an acausal connecting principle" was a provocative hypothesis when it first was published and has remained so up to the present. In it C. G. Jung aimed at expanding the Western world's core conceptions of nature and the psyche. By requiring that we include and make room for unique individual experiences of life in our most fundamental philosophical and scientific views of the world, Jung challenged the status quo, urging us to go beyond the readily explainable, beyond the restrictions of a cause-effect reductive description of the world, to seeing the psyche as embedded into the substance of the world.

Carl Jung in 1910

Jung's theories, practices and clinical methods bear direct relationship to what currently is referred to as *complexity theory*.

Acknowledgment

As an instance of complementarity, I credit/blame mathematician Donald Wehn, a fan of Jung, for enticing/burdening me to write this foreword. Donald wrote that "The issue which brought Pauli and Jung together was their shared interest in Reality.... Kepler's view contradicted many hundreds of years of church views. What was being lost by his new view of planetary reality? We will have to wait to read Pauli's essay."

References

Cambray, Joseph. *Synchronicity: Nature & Psyche in an Interconnected Universe*, TAMU Press, 2012

Enz, Charles P. *No Time to be Brief: A Scientific Biography of Wolfgang Pauli*, Oxford 2010.

Lindorff, David. *Pauli and Jung: The Meeting of Two Great Minds*, Quest Books, 2004.

Loehrich, Rolf R. *Exercitium Cogitandi*, six volumes (vol.2 with Lawrence G. Roberts), Oxford: Centre for Medieval & Renaissance Studies, 1978.

Miller, Arthur I. *137: Jung, Pauli, and the Pursuit of a Scientific Obsession*, Norton 2010.

Wehn, Donald. *Probabilities on Lie Groups*, Proceedings of the National Academy of Sciences, Vol. 48, 1962, 791–795.

Email for Marvin Jay Greenberg: mjg0@pacbell.net .

CONTENTS

LIST OF PLATES

ACKNOWLEDGMENTS

THE PUBLISHERS wish to express gratitude for permission to quote, as follows: to the Harvard University Press and William Heinemann, Ltd., London, for a passage from the Loeb Classical Library edition of Philo; to E. P. Dutton and Co., New York, and J. M. Dent and Sons, Ltd., London, for quotations from *Leibniz: Philosophical Writings,* in Everyman's Library; to Houghton, Mifflin Co., Boston, and Allen and Unwin, Ltd., London, for passages from Arthur Waley's *The Way and Its Power;* and to Zeno publishers, London, for a passage from John Precope's edition of Hippocrates.

C. G. JUNG

Synchronicity:
An Acausal Connecting Principle

Translated from the German by

R. F. C. HULL

EDITORIAL NOTE

THIS WORK is a translation of "Synchronizität als ein Prinzip akausaler Zusammenhänge," which, together with Professor Pauli's monograph, composed the volume *Naturerklärung und Psyche,* a publication of the C. G. Jung Institute in Zurich. For the English edition, the text has been revised by the author; material has been added or rearranged, but the essential argument remains unaltered.

This version has been prepared with the editorial collaboration of the Editors of the Collected Works of C. G. Jung. It will later take its place in the Collected Works as part of the contents of Volume 8, *The Structure and Dynamics of the Psyche.* Part of the material originally was presented as a brief essay, "Über Synchronizität," in *Eranos-Jahrbuch 1951,* which will be published in English as "On Synchronicity" in *Man and Time* (Papers from the Eranos Yearbooks, 3).

TRANSLATOR'S NOTE

I WANT to express my thanks to the following persons for their help and advice: Professor Max Knoll, for assistance with certain technical terminology; Professor Erwin Panofsky, for translating passages from Kepler; Dr. Marie-Louise von Franz, for translating passages from Latin and Greek; Norbert Guterman, for assistance with a passage from Paracelsus; A. S. B. Glover, for bibliographical and other valuable advice; Miss Barbara Hannah, for her meticulous work on the typescript; and Professor Jung himself, for reviewing and correcting the typescript and proof and patiently answering a multitude of queries.

FOREWORD

In writing this paper I have, so to speak, made good a promise which for many years I lacked the courage to fulfil. The difficulties of the problem and its presentation seemed to me too great; too great the intellectual responsibility without which such a subject cannot be tackled; too inadequate, in the long run, my scientific training. If I have now conquered my hesitation and at last come to grips with my theme, it is chiefly because my experiences of the phenomenon of synchronicity have multiplied themselves over the decades, while on the other hand my researches into the history of symbols, and of the fish symbol in particular, brought the problem ever closer to me, and finally because I have been alluding to the existence of this phenomenon on and off in my writings for twenty years without discussing it any further. I would like to put a temporary end to this unsatisfactory state of affairs by trying to give a consistent account of everything I have to say on this subject. I hope it will not be construed as presumption on my part if I make uncommon demands on the open-mindedness and goodwill of the reader. Not only is he expected to plunge into regions of human experience which are dark, dubious, and hedged about with prejudice, but the intellectual difficulties are such as the treatment and elucidation of so abstract a subject must inevitably entail. As anyone can see for himself after reading a few pages, there can be no question of a complete description and explanation of these complicated phenomena, but only an attempt to broach the problem in such a way as to reveal some of its manifold aspects and connections, and to open up a very obscure field which is philosophically of the greatest importance. As a psychiatrist and psychotherapist I have often come up against the phenomena in question and could convince myself how much these inner experiences meant to my patients. In most cases they were things which people do not talk about for fear of exposing themselves to thoughtless ridicule. I was amazed to see how many people have had experiences of this kind and how carefully the secret was guarded. So my interest in this problem has a human as well as a scientific foundation.

In the performance of my work I had the support of a number of friends who are mentioned in the text. Here I would like to express my particular thanks to Dr. Liliane Frey-Rohn, for her help with the astrological material.

August, 1950 C. G. JUNG

CHAPTER ONE

EXPOSITION

The discoveries of modern physics have, as we know, brought about a significant change in our scientific picture of the world, in that they have shattered the absolute validity of natural law and made it relative. Natural laws are *statistical* truths, which means that they are completely valid only when we are dealing with macrophysical quantities. In the realm of very small quantities *prediction* becomes uncertain, if not impossible, because very small quantities no longer behave in accordance with the known natural laws.

The philosophical principle that underlies our conception of natural law is *causality*. But if the connection between cause and effect turns out to be only statistically valid and only relatively true, then the causal principle is only of relative use for explaining natural processes and therefore presupposes the existence of one or more other factors which would be necessary for an explanation. This is as much as to say that the connection of events may in certain circumstances be other than causal, and requires another principle of explanation.[1]

[1] [Other than, or supplementary to, the laws of chance.—EDS.]

We shall naturally look round in vain in the macrophysical world for acausal events, for the simple reason that we cannot imagine events that are connected non-causally and are capable of a non-causal explanation. But that does not mean that such events do not exist. Their existence—or at least their possibility—follows logically from the premise of statistical truth.

The experimental method of inquiry aims at establishing regular events which can be repeated. Consequently, unique or rare events are ruled out of account. Moreover, the experiment imposes limiting conditions on nature, for its aim is to force her to give answers to questions devised by man. Every answer of nature is therefore more or less influenced by the kind of questions asked, and the result is always a hybrid product. The so-called "scientific view of the world" [2] based on this can hardly be anything more than a psychologically biased partial view which misses out all those by no means unimportant aspects that cannot be grasped statistically. But, to grasp these unique or rare events at all, we seem to be dependent on equally "unique" and individual descriptions. This would result in a chaotic collection of curiosities, rather like those old natural history cabinets where one finds, cheek by jowl with fossils and anatomical monsters in bottles, the horn of a unicorn, a mandragora manikin, and a dried mermaid. The descriptive sciences, and above all biology in the widest sense, are familiar with these "unique" specimens, and in their case only *one* example of an organism, no matter how unbelievable it may be, is needed to establish its existence. At any rate numerous observers will be able

[2] ["naturwissenschaftliche Weltanschauung."]

to convince themselves, on the evidence of their own eyes, that such a creature does in fact exist. But where we are dealing with ephemeral events which leave no demonstrable traces behind them except fragmentary memories in people's minds, then a single witness no longer suffices, nor would several witnesses be enough to make a unique event appear absolutely credible. One has only to think of the notorious unreliability of eye-witness accounts. In these circumstances we are faced with the necessity of finding out whether the apparently unique event is really unique in our recorded experience, or whether the same or similar events are not to be found elsewhere. Here the *consensus omnium* plays a very important role psychologically, though empirically it is somewhat doubtful, for only in exceptional cases does the *consensus omnium* prove to be of value in establishing facts. The empiricist will not leave it out of account, but will do better not to rely on it. Absolutely unique and ephemeral events whose existence we have no means of either denying or proving can never be the object of empirical science; rare events might very well be, provided that there was a sufficient number of reliable individual observations. The so-called *possibility* of such events is of no importance whatever, for the criterion of what is possible in any age is derived from that age's rationalistic assumptions. There are no "absolute" natural laws to whose authority one can appeal in support of one's prejudices. The most that can fairly be demanded is that the number of individual observations shall be as high as possible. If this number, statistically considered, falls within the limits of chance expectation, then it has been statistically proved that it was a question of chance; but no *explanation* has thereby been furnished. There has merely been an exception to the rule. When, for instance, the number of symptoms indicating a complex falls below the probable number of disturbances to be expected during the association experiment, this is no justification for assuming that no complex exists. But that did not prevent the reaction disturbances from being regarded earlier as pure chance.[3]

Although, in biology especially, we move in a sphere where causal explanations often seem very unsatisfactory —indeed, well-nigh impossible—we shall not concern ourselves here with the problems of biology, but rather with the question whether there may not be some general field where acausal events not only are possible but are found to be actual facts.

Now, there is in our experience an immeasurably wide field whose extent forms, as it were, the counterbalance to the domain of causality. This is the world of chance, where a chance event seems causally unconnected with the coinciding fact. So we shall have to examine the nature and the whole idea of chance a little more closely. Chance, we say, must obviously be susceptible of some causal explanation and is only called "chance" or "coincidence" because its causality has not yet been discovered. Since we have an inveterate conviction of the absolute validity of causal law, we regard this explanation of chance as being quite adequate. But if the causal principle is only relatively valid, then it follows that even though in the vast majority of cases an apparently chance series can be

[3] [Cf. Jung, *Studies in Word Association,* trans. by M. D. Eder (London, 1918; New York, 1919).—EDS.]

causally explained, there must still remain a number of cases which do not show causal connection. We are therefore faced with the task of sifting events and separating the acausal ones from those that can be causally explained. It stands to reason that the number of causally explicable events will far exceed those suspected of acausality, for which reason a superficial or prejudiced observer may easily overlook the relatively rare acausal phenomena. As soon as we come to deal with the problem of chance the need for a statistical evaluation of the events in question forces itself upon us.

It is not possible to sift the empirical material without a criterion of distinction. How are we to recognize acausal combinations of events, since it is obviously impossible to examine all chance happenings for their causality? The answer to this is that acausal events may be expected most readily where, on closer reflection, a causal connection appears to be inconceivable. As an example I would cite the "duplication of cases" which is a phenomenon well known to every doctor. Occasionally there is a trebling or even more, so that Kammerer [4] can speak of a "law of series," of which he gives a number of excellent examples. In the majority of such cases there is not even the remotest probability of a causal connection between the coinciding events. When for instance I am faced with the fact that my tram ticket bears the same number as the theatre ticket which I buy immediately afterwards, and I receive that same evening a telephone call during

[4] Paul Kammerer, *Das Gesetz der Serie* (Stuttgart and Berlin, 1919).

which the same number is mentioned again as a telephone number, then a causal connection between them seems to me improbable in the extreme, although it is obvious that each event must have its own causality. I know, on the other hand, that chance happenings have a tendency to fall into aperiodic groupings—necessarily so, because otherwise there would be only a periodic or regular arrangement of events which would by definition exclude chance.

Kammerer holds that, though "runs" [5] or successions of chance events are not subject to the operation of a common cause,[6] i.e., are acausal, they are nevertheless an expression of inertia—the property of persistence.[7] The simultaneity of a "run of the same thing side by side" he explains as "imitation." [8] Here he contradicts himself, for the run of chance has not been "removed outside the realm of the explicable," [9] but, as we would expect, remains inside it and is consequently reducible, if not to a common cause, then at least to several causes. His concepts of seriality, imitation, attraction, and inertia belong to a causally conceived view of the world and tell us no more than that the run of chance corresponds to statisti-

[5] Ibid., p. 130.

[6] Pp. 36, 93 *f*, 102 *f*.

[7] "The law of series is an expression of the inertia of the objects involved in its repetitions (i.e., producing the series). The far greater inertia of a complex of objects and forces (as compared to that of a single object or force) explains the persistence of an identical constellation and the emergence, connected therewith, of repetitions over long periods of time" (p. 117).

[8] P. 130.

[9] P. 94.

cal and mathematical probability.[10] Kammerer's factual material contains nothing but runs of chance whose only "law" is probability; in other words, there is no apparent reason why he should look behind them for anything else. But for some obscure reason he does look behind them for something more than mere probability warrants—for a *law of seriality* which he would like to introduce as a principle coexistent with causality and finality. This tendency, as I have said, is in no way justified by his material. I can only explain this obvious contradiction by supposing that he had a dim but fascinated intuition of an acausal arrangement and combination of events, probably because, like all thoughtful and sensitive natures, he could not escape the peculiar impression which runs of chance usually make on us, and therefore, in accordance with his scientific disposition, took the bold step of postulating an acausal seriality on the basis of empirical material that lay within the limits of probability. Unfortunately he did not attempt a quantitative evaluation of seriality. Such an undertaking would undoubtedly have thrown up questions that are difficult to answer. The casuistic method serves well enough for the purpose of general orientation, but only quantitative evaluation or the statistical method promises results in dealing with chance.

Chance groupings or series seem, at least to our present way of thinking, to be meaningless, and to fall as a general rule within the limits of probability. There are, however, incidents whose "chancefulness" seems open to

[10] [The term "probability" therefore refers to the probability on a chance hypothesis (Null Hypothesis). This is the sense in which the term is usually used in this paper.—EDS.]

doubt. To mention but one example out of many, I noted the following on April 1, 1949: Today is Friday. We have fish for lunch. Somebody happens to mention the custom of making an "April fish" of someone. That same morning I made a note of an inscription which read: "Est homo totus medius *piscis* ab imo." In the afternoon a former patient of mine, whom I had not seen in months, showed me some extremely impressive pictures of fish which she had painted in the meantime. In the evening I was shown a piece of embroidery with fish-like sea-monsters in it. On the morning of April 2 another patient, whom I had not seen for many years, told me a dream in which she stood on the shore of a lake and saw a large fish that swam straight towards her and landed at her feet. I was at this time engaged on a study of the fish symbol in history. Only one of the persons mentioned here knew anything about it.

The suspicion that this must be a case of *meaningful coincidence,* i.e., an acausal connection, is very natural. I must own that this run of events made a considerable impression on me. It seemed to me to have a certain numinous quality.[11] In such circumstances we are inclined to say, "That cannot be mere chance," without knowing what exactly we are saying. Kammerer would no doubt have reminded me of his "seriality." The strength of an

[11] The numinosity of a series of chance happenings grows in proportion to the number of its terms. Unconscious—probably archetypal—contents are thereby constellated, which then give rise to the impression that the series has been "caused" by these contents. Since we cannot conceive how this could be possible without recourse to positively magical categories, we generally let it go at the bare impression.

impression, however, proves nothing against the fortuitous coincidence of all these fishes. It is, admittedly, exceedingly odd that the fish theme recurs no less than six times within twenty-four hours. But one must remember that fish on Friday is the usual thing, and on April 1 one might very easily think of the April fish. I had at that time been working on the fish symbol for several months. Fishes frequently occur as symbols of unconscious contents. So there is no possible justification for seeing in this anything but a chance grouping. Runs or series which are composed of quite ordinary occurrences must for the present be regarded as fortuitous.[12] However wide their range may be, they must be ruled out as acausal connections. It is, therefore, generally assumed that all coincidences are lucky hits and do not require an acausal interpretation.[13]

[12] As a pendant to what I have said above, I should like to mention that I wrote these lines sitting by the lake. Just as I had finished this sentence, I walked over to the sea-wall and there lay a dead fish, about a foot long, apparently uninjured. No fish had been there the previous evening. (Presumably it had been pulled out of the water by a bird of prey or a cat.) The fish was the seventh in the series.

[13] We find ourselves in something of a quandary when it comes to making up our minds about the phenomenon which Stekel calls the "compulsion of the name." What he means by this is the sometimes quite grotesque coincidence between a man's name and his peculiarities or profession. For instance Herr Gross (Mr. Grand) suffers from delusions of grandeur, Herr Kleiner (Mr. Small) has an inferiority complex. The Altmann sisters marry men twenty years older than themselves. Herr Feist (Mr. Stout) is the Food Minister, Herr Rosstäuscher (Mr. Horsetrader) is a lawyer, Herr Kalberer (Mr. Calver) is an obstetrician, Herr Freud (joy) champions the pleasure-principle, Herr Adler (eagle) the will-to-power, Herr Jung (young) the idea of rebirth, and so on. Are these the whimsicalities of chance, or the suggestive

This assumption can, and indeed must, be regarded as true so long as proof is lacking that their incidence exceeds the limits of probability. Should this proof be forthcoming, however, it would prove at the same time that there are genuinely non-causal combinations of events for whose explanation we should have to postulate a factor incommensurable with causality. We should then have to assume that events in general are related to one another on the one hand as causal chains, and on the other hand by a kind of *meaningful cross-connection.*

Here I should like to draw attention to a treatise of Schopenhauer's, "On the Apparent Design in the Fate of the Individual," [14] which originally stood godfather to the views I am now developing. It deals with the "simultaneity of the causally unconnected, which we call 'chance.' " [15] Schopenhauer illustrates this simultaneity by a geographical analogy, where the parallels represent the cross-connection between the meridians, which are thought of as causal chains.[16]

> All the events in a man's life would accordingly stand in two fundamentally different kinds of connection: firstly, in the objective, causal connection of the natural process; secondly, in a subjective connection which exists only in relation to the individual who experiences it, and which is thus as subjective as his own dreams.

effects of the name, as Stekel seems to suggest, or are they "meaningful coincidences"? ("Die Verpflichtung des Namens," *Zeitschrift für Psychotherapie und medizinische Psychologie,* Stuttgart, III, 1911, 110 *ff.*)

[14] *Parerga und Paralipomena,* I, ed. by R. von Koeber (Berlin, 1891). [Cf. the trans. by David Irvine (London, 1913), to which reference is made for convenience, though not quoted here.]

[15] Ibid., p. 40. [Irvine, p. 41.]

[16] P. 39. [Irvine, pp. 39 *f.*]

. . . That both kinds of connection exist simultaneously, and the selfsame event, although a link in two totally different chains, nevertheless falls into place in both, so that the fate of one individual invariably fits the fate of the other, and each is the hero of his own drama while simultaneously figuring in a drama foreign to him—this is something that surpasses our powers of comprehension and can only be conceived as possible by virtue of the most wonderful pre-established harmony.[17]

In his view "the subject of the great dream of life . . . is but one,"[18] the transcendental Will, the *prima causa,* from which all causal chains radiate like meridian lines from the poles and, because of the circular parallels, stand to one another in a meaningful relationship of simultaneity.[19] Schopenhauer believed in the absolute determinism of the natural process and furthermore in a first cause. There is nothing to warrant either assumption. The first cause is a philosophical mythologem which is only credible when it appears in the form of the old paradox Ἓν τὸ πᾶν, as unity and multiplicity at once. The idea that the simultaneous points in the causal chains, or meridians, represent meaningful coincidences would only hold water if the first cause really were a unity. But if it were a multiplicity, which is just as likely, then Schopenhauer's whole explanation collapses, quite apart from the fact, which we have only recently realized, that natural law possesses a merely statistical validity and thus keeps the door open to indeterminism. Neither philosophical reflection nor experience can provide any evidence for the regular existence of these two kinds of connection, in which the same thing is both subject and object. Schopenhauer

[17] P. 45. [Irvine, pp. 49 *f.*]

[18] P. 46. [Irvine, p. 50.]

[19] Hence my term "synchronicity."

thought and wrote at a time when causality held sovereign sway as a category *a priori* and had therefore to be dragged in to explain meaningful coincidences. But, as we have seen, it can do this with some degree of probability only if we have recourse to the other, equally arbitrary assumption of the unity of the first cause. It then follows as a *necessity* that every point on a given meridian stands in a relationship of meaningful coincidence to every other point on the same degree of latitude. This conclusion, however, goes far beyond the bounds of what is empirically possible, for it credits meaningful coincidences with occurring so regularly and systematically that their verification would be either unnecessary or the simplest thing in the world. Schopenhauer's examples carry as much or as little conviction as all the others. Nevertheless, it is to his credit that he saw the problem and understood that there are no facile *ad hoc* explanations. Since this problem is concerned with the foundations of our epistemology, he derived it in accordance with the general trend of his philosophy from a transcendental premise, from the Will which creates life and being on all levels, and which modulates each of these levels in such a way that they are not only in harmony with their synchronous parallels but also prepare and arrange future events in the form of Fate or Providence.

In contrast to Schopenhauer's accustomed pessimism, this utterance has an almost friendly and optimistic tone which we can hardly sympathize with today. One of the most problematical and momentous centuries the world has ever known separates us from that still medievalistic age when the philosophizing mind believed it could make assertions beyond what could be empirically proved. It was an age of large views, which did not cry halt and think that the limits of nature had been reached just where the scientific road-builders had come to a temporary stop. Thus Schopenhauer, with true philosophical vision, opened up a field for reflection whose peculiar phenomenology he was not equipped to understand, though he outlined it more or less correctly. He recognized that with their *omina* and *praesagia* astrology and the various intuitive methods of interpreting fate have a common denominator which he sought to discover by means of "transcendental speculation." He recognized, equally rightly, that it was a problem of principle of the first order, unlike all those before and after him who operated with futile conceptions of some kind of energy transmission, or conveniently dismissed the whole thing as nonsense in order to avoid a too difficult task.[20] Schopenhauer's attempt is the more remarkable in that it was made at a time when the tremendous advance of the natural sciences had convinced everybody that causality alone could be considered the final principle of explanation. Instead of ignoring all those experiences which refuse to bow down to the sovereign rule of causality, he tried, as we have seen, to fit them into his deterministic view of the world. In so doing, he forced concepts like prefiguration, correspondence, and pre-established harmony, which as a universal order coexisting with the causal one have always underlain man's explanations of nature, into the causal scheme, probably because he

[20] Here I must make an exception of Kant, whose treatise "Dreams of a Spirit-Seer, Illustrated by Dreams of Metaphysics" (English trans., London, 1900) pointed the way for Schopenhauer.

felt—and rightly—that the scientific view of the world based on natural law, though he did not doubt its validity, nevertheless lacked something which played a considerable role in the classical and medieval view (as it also does in the intuitive feelings of modern man).

The mass of facts collected by Gurney, Myers, and Podmore [21] inspired three other investigators—Dariex,[22] Richet,[23] and Flammarion [24]—to tackle the problem in terms of a probability calculus. Dariex found a probability of 1 : 4,114,545 for telepathic precognitions of death, which means that the explanation of such a warning as due to "chance" is more than four million times more improbable than explaining it as a "telepathic," or acausal, meaningful coincidence. The astronomer Flammarion reckoned a probability of no less than 1 : 804,622,222 for a particularly well-observed instance of "phantasms of the living." [25] He was also the first to link up other suspicious happenings with the general interest in phenomena connected with death. Thus he relates [26] that, while writing his book on the atmosphere, he was just at the chapter on wind-force when a sudden gust of wind swept all his

[21] Edmund Gurney, Frederic W. H. Myers, and Frank Podmore, *Phantasms of the Living* (2 vols., London, 1886).

[22] Xavier Dariex, "Le Hazard et la Télépathie," *Annales des sciences psychiques* (Paris), I (1891), 295–304.

[23] Charles Richet, "Relations de diverses expériences sur transmission mentale, la lucidité, et autres phénomènes non explicable par les données scientifiques actuelles," *Proceedings of the Society for Psychical Research* (London), V (1888), 18–168.

[24] Camille Flammarion, *The Unknown* (London and New York, 1900), pp. 191 *ff.*

[25] Ibid., p. 202.

[26] Pp. 192 *f.*

papers off the table and blew them out of the window. He also cites, as an example of triple coincidence, the edifying story of Monsieur de Fortgibu and the plum-pudding.[27] The fact that he mentions these coincidences at all in connection with the problem of telepathy shows that Flammarion had a distinct intuition, albeit an unconscious one, of a far more comprehensive principle.

The writer Wilhelm von Scholz [28] has collected a number of stories showing the strange ways in which lost or stolen objects come back to their owners. Among other things, he tells the story of a mother who took a photograph of her small son in the Black Forest. She left the film to be developed in Strasbourg. But, owing to the outbreak of war, she was unable to fetch it and gave it up for lost. In 1916 she bought a film in Frankfort in order to take a photograph of her daughter, who had been born in the meantime. When the film was developed it was found to be doubly exposed: the picture underneath was the photograph she had taken of her son in 1914! The old film had not been developed and had somehow got into circulation again among the new films. The author

[27] Pp. 194 *ff*. A certain M. Deschamps, when a boy in Orléans, was once given a piece of plum-pudding by a M. de Fortgibu. Ten years later he discovered another plum-pudding in a Paris restaurant, and asked if he could have a piece. It turned out, however, that the plum-pudding was already ordered—by M. de Fortgibu. Many years afterwards M. Deschamps was invited to partake of a plum-pudding as a special rarity. While he was eating it he remarked that the only thing lacking was M. de Fortgibu. At that moment the door opened and an old, old man in the last stages of disorientation walked in: M. de Fortgibu, who had got hold of the wrong address and burst in on the party by mistake.

[28] *Der Zufall: Eine Vorform des Schicksals* (Stuttgart, 1924).

comes to the understandable conclusion that everything points to the "mutual attraction of related objects," or an "elective affinity." He suspects that these happenings are arranged as if they were the dream of a "greater and more comprehensive consciousness, which is unknowable."

The problem of chance has been approached from the psychological angle by Herbert Silberer.[29] He shows that apparently meaningful coincidences are partly unconscious arrangements, and partly unconscious, arbitrary interpretations. He takes no account either of parapsychic phenomena or of synchronicity, and theoretically he does not go much beyond the causalism of Schopenhauer. Apart from its valuable psychological criticism of our methods of evaluating chance, Silberer's study contains no reference to the occurrence of meaningful coincidences as here understood.

Decisive evidence for the existence of acausal combinations of events has been furnished, with adequate scientific safeguards, only very recently, mainly through the experiments of J. B. Rhine and his fellow-workers,[30] who

[29] *Der Zufall und die Koboldstreiche des Unbewussten* (Schriften zur Seelenkunde und Erziehungskunst, No. III; Bern and Leipzig, 1921).

[30] J. B. Rhine, *Extra-Sensory Perception* (Boston, 1934) and *New Frontiers of the Mind* (New York, 1937). J. G. Pratt, J. B. Rhine, C. E. Stuart, B. M. Smith, and J. A. Greenwood, *Extra-Sensory Perception after Sixty Years* (New York, 1940). A general survey of the findings in Rhine, *The Reach of the Mind* (London and New York, 1948; in Penguin Books, 1954), and also in the valuable book by G. N. M. Tyrrell, *The Personality of Man* (Penguin Books, London, 1947). A short résumé in Rhine, "An Introduction to the Work of Extra-Sensory Perception," *Transactions of the New York Academy of Sciences,* Series II, XII (1950),

have not, however, recognized the far-reaching conclusions that must be drawn from their findings. Up to the present no critical argument that cannot be refuted has been brought against these experiments. The experiment consists, in principle, in an experimenter turning up, one after another, a series of numbered cards bearing simple geometrical patterns. At the same time the subject, separated by a screen from the experimenter, is given the task of guessing the signs as they are turned up. A pack of twenty-five cards is used, each five of which carry the same sign. Five cards are marked with a star, five with a square, five with a circle, five with wavy lines, and five with a cross. The experimenter naturally does not know the order in which the pack is arranged, nor has the subject any opportunity of seeing the cards. Many of the experiments were negative, since the result did not exceed the probability of five chance hits. In the case of certain subjects, however, some results were distinctly above probability. The first series of experiments consisted in each subject trying to guess the cards 800 times. The average result showed 6.5 hits for 25 cards, which is 1.5 more than the chance probability of 5 hits. The probability of there being a chance deviation of 1.5 from the number 5 works out at 1 : 250,000. This proportion shows that the probability of a chance deviation is not exactly high, since it is to be expected only once in 250,000 cases. The results vary according to the specific gift of the individual subject. One young man, who in numerous experiments scored an average of 10 hits for every 25 cards (double the probable

164 *ff.* S. G. Soal and F. Bateman, *Modern Experiments in Telepathy* (London, 1954).

number), once guessed all 25 cards correctly, which gives a probability of 1 : 298,023,223,876,953,125. The possibility of the pack being shuffled in some arbitrary way is guarded against by an apparatus which shuffles the cards automatically, independently of the experimenter.

After the first series of experiments the spatial distance between the experimenter and the subject was increased, in one case to 250 miles. The average result of numerous experiments amounted here to 10.1 hits for 25 cards. In another series of experiments, when experimenter and subject were in the same room, the score was 11.4 for 25; when the subject was in the next room, 9.7 for 25; when two rooms away, 12.0 for 25. Rhine mentions the experiments of F. L. Usher and E. L. Burt, which were conducted with positive results over a distance of 960 miles.[31] With the aid of synchronized watches experiments were also conducted between Durham, North Carolina, and Zagreb, Yugoslavia, about 4000 miles, with equally positive results.[32]

The fact that distance has no effect in principle shows that the thing in question cannot be a phenomenon of force or energy, for otherwise the distance to be overcome and the diffusion in space would cause a diminution of the effect, and it is more than probable that the score would fall proportionately to the square of the distance. Since this is obviously not the case, we have no alternative but to assume that distance is psychically

[31] *The Reach of the Mind* (1954 edn.), p. 48.

[32] Rhine and Betty M. Humphrey, "A Transoceanic ESP Experiment," *The Journal of Parapsychology* (Durham), VI (1942), 52 *ff.*

variable, and may in certain circumstances be reduced to vanishing point by a psychic condition.

Even more remarkable is the fact that *time* is not in principle a prohibiting factor either; that is to say, the reading of a series of cards to be turned up in the future produces a score that exceeds chance probability. The results of Rhine's time experiment show a probability of 1 : 400,000, which means a considerable probability of there being some factor independent of time. They point, in other words, to a psychic relativity of time, since the experiment was concerned with perceptions of events which had not yet occurred. In these circumstances the time factor seems to have been eliminated by a psychic function or psychic condition which is also capable of abolishing the spatial factor. If, in the spatial experiments, we were obliged to admit that energy does not decrease with distance, then the time experiments make it completely impossible for us even to think of there being any energy relationship between the perception and the future event. We must give up at the outset all explanations in terms of energy, which amounts to saying that events of this kind cannot be considered from the point of view of causality, for causality presupposes the existence of space and time in so far as all observations are ultimately based upon bodies in motion.

Among Rhine's experiments we must also mention the experiments with dice. The subject has the task of throwing the dice (which is done by an apparatus), and at the same time he has to wish that one number (say 3) will turn up as many times as possible. The results of this so-called PK (psychokinetic) experiment were positive, the more so the more dice were used at one time.[33] If space and time prove to be psychically relative, then the moving body must possess, or be subject to, a corresponding relativity.

One consistent experience in all these experiments is the fact that the number of hits scored tends to sink after the first attempt, and the results then become negative. But if, for some inner or outer reason, there is a freshening of interest on the subject's part, the score rises again. Lack of interest and boredom are negative factors; enthusiasm, positive expectation, hope, and belief in the possibility of ESP make for good results and seem to be the real conditions which determine whether there are going to be any results at all. In this connection it is interesting to note that the well-known English medium, Mrs. Eileen J. Garrett, achieved bad results in the Rhine experiments because, as she herself admits, she was unable to summon up any feeling for the "soulless" test-cards.

These few hints may suffice to give the reader at least a superficial idea of these experiments. The above-mentioned book by G. N. M. Tyrrell, late president of the Society for Psychical Research, contains an excellent summing-up of all experiences in this field. Its author himself rendered great service to ESP research. From the physicist's side the ESP experiments have been evaluated in a positive sense by Robert A. McConnell in an article entitled "ESP—Fact or Fancy?"[34]

[33] *The Reach of the Mind,* pp. 75 *ff.*

[34] Professor Pauli was kind enough to draw my attention to this paper, which appeared in *The Scientific Monthly* (London), LXIX (1949), No. 2.

As is only to be expected, every conceivable kind of attempt has been made to explain away these results, which seem to border on the miraculous and frankly impossible. But all such attempts come to grief on the facts, and the facts refuse so far to be argued out of existence. Rhine's experiments confront us with the fact that there are events which are related to one another experimentally, and in this case *meaningfully,* without there being any possibility of proving that this relation is a causal one, since the "transmission" exhibits none of the known properties of energy. There is therefore good reason to doubt whether it is a question of transmission at all. The time experiments rule out any such thing in principle, for it would be absurd to suppose that a situation which does not yet exist and will only occur in the future could transmit itself as a phenomenon of energy to a receiver in the present.[35] It seems more likely that scientific explanation will have to begin with a criticism of our concepts of space and time on the one hand, and with the unconscious on the other. As I have said, it is impossible, with our present resources, to explain ESP, or the fact of meaningful coincidence, as a phenomenon of energy. This makes an end of the causal explanation as well, for "effect" cannot be understood as anything except a phenomenon of energy. Therefore it cannot be a question of cause and effect, but of a falling together in time, a kind of simultaneity. Because of this quality of simultaneity, I have picked on the term "synchronicity" to designate a hypothetical factor

[35] Kammerer has dealt, not altogether convincingly, with the question of the "countereffect of the succeeding state on the preceding one" (cf. *Das Gesetz der Serie,* pp. 131 *f.*).

equal in rank to causality as a principle of explanation. In my essay "The Spirit of Psychology" [36] I defined synchronicity as a psychically conditioned relativity of space and time. Rhine's experiments show that in relation to the psyche space and time are, so to speak, "elastic" and can apparently be reduced almost to vanishing point, as though they were dependent on psychic conditions and did not exist in themselves but were only "postulated" by the conscious mind. In man's original view of the world, as we find it among primitives, space and time have a very precarious existence. They became "fixed" concepts only in the course of his mental development, thanks largely to the introduction of measurement. In themselves, space and time consist of *nothing*. They are hypostatized concepts born of the discriminating activity of the conscious mind, and they form the indispensable co-ordinate for describing the behaviour of bodies in motion. They are, therefore, essentially psychic in origin, which is probably the reason that impelled Kant to regard them as *a priori* categories. But if space and time are only apparently properties of bodies in motion and are created by the intellectual needs of the observer, then their relativization by psychic conditions is no longer a matter for astonishment but is brought within the bounds of possibility. This possibility presents itself when the psyche observes, not external bodies, but *itself*. That is precisely what happens in Rhine's experiments: the subject's answer is not the result of his observing the physical cards, it is a product of pure imagination, of "chance" ideas which reveal the structure of that

[36] Trans. in *Spirit and Nature* (Papers from the Eranos Yearbooks, 1; New York, 1954; London, 1955).

which produces them, namely the unconscious. Here I will only point out that it is the decisive factors in the unconscious psyche, the archetypes, which constitute the structure of the collective unconscious. The latter represents a psyche that is identical with itself in all individuals. It cannot be directly perceived or "represented," in contrast to the perceptible psychic phenomena, and on account of its "irrepresentable" nature I have called it "psychoid."

The archetypes are formal factors responsible for the organization of unconscious psychic processes: they are "patterns of behaviour." At the same time they have a "specific charge" and develop numinous effects which express themselves as *affects*. The affect produces a partial *abaissement du niveau mental,* for although it raises a particular content to a supernormal degree of luminosity, it does so by withdrawing so much energy from other possible contents of consciousness that they become darkened and eventually unconscious. Owing to the restriction of consciousness produced by the affect so long as it lasts, there is a corresponding lowering of orientation which in its turn gives the unconscious a favourable opportunity to slip into the space vacated. Thus we regularly find that unexpected or otherwise inhibited unconscious contents break through and find expression in the affect. Such contents are very often of an inferior or primitive nature and thus betray their archetypal origin. As I shall show further on, certain phenomena of simultaneity or synchronicity seem to be bound up with the archetypes.

The extraordinary spatial orientation of animals may also point to the psychic relativity of space and time. The puzzling time-orientation of the palolo worm, for instance, whose tail-segments, loaded with sexual products, always appear on the surface of the sea the day before the last quarter of the moon in October and November,[37] might be mentioned in this connection. One of the causes suggested is the acceleration of the earth owing to the gravitational pull of the moon at this time. But, for astronomical reasons, this explanation cannot possibly be right.[38] The relation which undoubtedly exists between the human menstruation period and the course of the moon is connected with the latter only numerically and does not really coincide with it. Nor has it been proved that it ever did.

The problem of synchronicity has puzzled me for a long time, ever since the middle twenties,[39] when I was investigating the phenomena of the collective unconscious

[37] To be more accurate, the swarming begins a little before and ends a little after this day, when the swarming is at its height. The months vary according to location. The palolo worm, or wawo, of Amboina is said to appear at full moon in March. (A. F. Krämer, *Über den Bau der Korallenriffe*, Kiel and Leipzig, 1897.)

[38] Fritz Dahns, "Das Schwärmen des Palolo," *Der Naturforscher* (Lichterfelde-Berlin), VIII (1932): 11, 379–82.

[39] Even before that time certain doubts had arisen in me as to the unlimited applicability of the causal principle in psychology. In the foreword to the 1st edition of *Collected Papers on Analytical Psychology,* ed. C. E. Long (London, 1916), I had written (p. xv): "Causality is only one principle and psychology essentially cannot be exhausted by causal methods only, because the mind [= psyche] lives by aims as well." Psychic finality rests on a "pre-existent" meaning which only becomes problematical when it is an unconscious arrangement. In that case we have to suppose a "knowledge" prior to all consciousness. Hans Driesch comes to the same conclusion (*Die "Seele" als elementarer Naturfaktor,* Leipzig, 1903, pp. 80 *ff.*).

and kept on coming across connections which I simply could not explain as chance groupings or "runs." What I found were "coincidences" which were connected so meaningfully that their "chance" concurrence would be incredible. By way of example, I shall mention an incident from my own observation. A young woman I was treating had, at a critical moment, a dream in which she was given a golden scarab. While she was telling me this dream I sat with my back to the closed window. Suddenly I heard a noise behind me, like a gentle tapping. I turned round and saw a flying insect knocking against the window pane from outside. I opened the window and caught the creature in the air as it flew in. It was the nearest analogy to a golden scarab that one finds in our latitudes, a scarabaeid beetle, the common rose-chafer (*Cetonia aurata*), which contrary to its usual habits had evidently felt an urge to get into a dark room at this particular moment. I must admit that nothing like it ever happened to me before or since, and that the dream of the patient has remained unique in my experience.

I should like to mention another case that is typical of a certain category of events. The wife of one of my patients, a man in his fifties, once told me in conversation that, at the deaths of her mother and her grandmother, a number of birds gathered outside the windows of the death-chamber. I had heard similar stories from other people. When her husband's treatment was nearing its end, his neurosis having been removed, he developed some apparently quite innocuous symptoms which seemed to me, however, to be those of heart-disease. I sent him along to a specialist, who after examining him told me in writing that he could find no cause for anxiety. On the way back from this consultation (with the medical report in his pocket) my patient collapsed in the street. As he was brought home dying, his wife was already in a great state of anxiety because, soon after her husband had gone to the doctor, a whole flock of birds alighted on their house. She naturally remembered the similar incidents that had happened at the death of her own relatives, and feared the worst.

Although I was personally acquainted with the people concerned and know very well that the facts here reported are true, I do not imagine for a moment that this will induce anybody who is determined to regard such things as pure "chance" to change his mind. My sole object in relating these two incidents is simply to give some indication of how meaningful coincidences usually present themselves in practical life. The meaningful connection is obvious enough in the first case in view of the approximate identity of the chief objects (the scarab and the beetle); but in the second case the death and the flock of birds seem to be incommensurable with one another. If one considers, however, that in the Babylonian Hades the souls wore a "feather dress," and that in ancient Egypt the *ba,* or soul, was thought of as a bird,[40] it is not too far-fetched to suppose that there may be some archetypal symbolism at work. Had such an incident occurred in a dream, that interpretation would be justified by the comparative psychological material. There also seems to be an archetypal foundation to the first case. It was an extraordinarily difficult case to treat, and up to the time of the

[40] In Homer the souls of the dead "twitter."

dream little or no progress had been made. I should explain that the main reason for this was my patient's animus, which was steeped in Cartesian philosophy and clung so rigidly to its own idea of reality that the efforts of three doctors—I was the third—had not been able to weaken it. Evidently something quite irrational was needed which was beyond my powers to produce. The dream alone was enough to disturb ever so slightly the rationalistic attitude of my patient. But when the "scarab" came flying in through the window in actual fact, her natural being could burst through the armour of her animus possession and the process of transformation could at last begin to move. Any essential change of attitude signifies a psychic renewal which is usually accompanied by symbols of rebirth in the patient's dreams and fantasies. The scarab is a classic example of a rebirth symbol. The ancient Egyptian Book of What Is in the Netherworld describes how the dead sun-god changes himself at the tenth station into Khepri, the scarab, and then, at the twelfth station, mounts the barge which carries the rejuvenated sun-god into the morning sky. The only difficulty here is that with educated people cryptomnesia often cannot be ruled out with certainty (although my patient did not happen to know this symbol). But this does not alter the fact that the psychologist is continually coming up against cases where the emergence of symbolic parallels [41] cannot be explained without the hypothesis of the collective unconscious.

Meaningful coincidences—which are to be distin-

[41] Naturally these can only be verified when the doctor himself has the necessary knowledge of symbology.

guished from meaningless chance groupings [42]—therefore seem to rest on an archetypal foundation. At least all the cases in my experience—and there is a large number of them—show this characteristic. What that means I have already suggested above.[43] Although anyone with any experience in this field can easily recognize their archetypal character, he will find it difficult to link them up with the psychic conditions in Rhine's experiments, because the latter contain no direct evidence of any constellation of the archetype. Nor is the emotional situation the same as in my examples. Nevertheless, it must be remembered that with Rhine the first series of experiments generally produced the best results, which then quickly fell off. But when it was possible to arouse a new interest in the essentially rather boring experiment, the results improved again. It follows from this that the emotional factor plays an important role. Affectivity, however, rests to a large extent on the instincts, whose formal aspect is the archetype.

There is yet another psychological analogy between my two cases and the Rhine experiments, though it is not quite so obvious. These apparently quite different situations have as their common characteristic an element of "impossibility." The patient with the scarab found herself

[42] [Statistical analysis is designed to separate out groupings (termed dispersions) due to random activity from significant dispersions in which causes may be looked for. On Professor Jung's hypothesis, however, dispersions due to chance can be subdivided into meaningful and meaningless. The meaningful dispersions due to chance are made meaningful by the activation of the psychoid archetype.—EDS.]

[43] Cf. "The Spirit of Psychology," *Spirit and Nature,* p. 416.

in an "impossible" situation because the treatment had got stuck and there seemed to be no way out of the impasse. In such situations, if they are serious enough, archetypal dreams are likely to occur which point out a possible line of advance one would never have thought of oneself. It is this kind of situation that constellates the archetype with the greatest regularity. In certain cases the psychotherapist therefore sees himself obliged to discover the rationally insoluble problem towards which the patient's unconscious is steering. Once this is found, the deeper layers of the unconscious, the primordial images, are activated and the transformation of the personality can get under way.

In the second case there was the half-unconscious fear and the threat of a lethal end with no possibility of an adequate recognition of the situation. In Rhine's experiment it is the "impossibility" of the task that ultimately fixes the subject's attention on the processes going on inside him, and thus gives the unconscious a chance to manifest itself. The questions set by the ESP experiment have an emotional effect right from the start, since they postulate something unknowable as being potentially knowable and in that way take the possibility of a miracle seriously into account. This, regardless of the subject's scepticism, immediately appeals to his unconscious readiness to witness a miracle, and to the hope, latent in all men, that such a thing may yet be possible. Primitive superstition lies just below the surface of even the most tough-minded individuals, and it is precisely those who most fight against it who are the first to succumb to its suggestive effects. When therefore a serious experiment with all the authority of science behind it touches this readiness, it will inevitably give rise to an emotion which either accepts or rejects it with a good deal of affectivity. At all events an affective expectation is present in one form or another even though it may be denied.

Here I would like to call attention to a possible misunderstanding which may be occasioned by the term "synchronicity." I chose this term because the simultaneous occurrence of two meaningfully but not causally connected events seemed to me an essential criterion. I am therefore using the general concept of synchronicity in the special sense of a coincidence in time of two or more causally unrelated events which have the same or a similar meaning, in contrast to "synchronism," which simply means the simultaneous occurrence of two events.

Synchronicity therefore means the simultaneous occurrence of a certain psychic state with one or more external events which appear as meaningful parallels to the momentary subjective state—and, in certain cases, vice versa. My two examples illustrate this in different ways. In the case of the scarab the simultaneity is immediately obvious, but not in the second example. It is true that the flock of birds occasioned a vague fear, but that can be explained causally. The wife of my patient was certainly not conscious beforehand of any fear that could be compared with my own apprehensions, for the symptoms (pains in the throat) were not of a kind to make the layman suspect anything bad. The unconscious, however, often knows more than the conscious, and it seems to me possible that the woman's unconscious had already got wind of the danger. If, therefore, we rule out a conscious psychic content such as the idea of deadly danger, there is an obvious simultaneity between the flock of birds, in its traditional meaning, and the death of the husband. The psychic state, if we disregard the possible but still not demonstrable excitation of the unconscious, appears to be dependent on the external event. The woman's psyche is nevertheless involved in so far as the birds settled on her house and were observed by her. For this reason it seems to me probable that her unconscious was in fact constellated. The flock of birds has, as such, a traditional mantic significance.[44] This is also apparent in the woman's own interpretation, and it therefore looks as if the birds represented an unconscious premonition of death. The physicians of the Romantic Age would probably have talked of "sympathy" or "magnetism." But, as I have said, such phenomena cannot be explained causally unless one permits oneself the most fantastic *ad hoc* hypotheses.

The interpretation of the birds as an omen is, as we have seen, based on two earlier coincidences of a similar kind. It did not yet exist at the time of the grandmother's death. There the coincidence was represented only by the death and the gathering of the birds. Both then and at the mother's death the coincidence was obvious, but in the third case it could only be verified when the dying man was brought into the house.

I mention these complications because they have an

[44] A literary example is "The Cranes of Ibycus." [A poem by Schiller (1798), inspired by the story of the Greek poet murdered by robbers who were brought to justice by cranes who saw the crime.—Eds.] Similarly, when a flock of chattering magpies settles on a house it is supposed to mean death, and so on. Cf. also the significance of auguries.

important bearing on the concept of synchronicity. Let us take another example: An acquaintance of mine saw and experienced in a dream the sudden death of a friend, with all the characteristic details. The dreamer was in Europe at the time and the friend in America. The death was confirmed next morning by telegram, and ten days later a letter confirmed the details. Comparison of European time with American time showed that the death occurred at least an hour before the dream. The dreamer had gone to bed late and not slept until about one o'clock. The dream occurred at approximately two in the morning. The dream experience is *not synchronous* with the death. Experiences of this kind frequently take place a little before or after the critical event. J. W. Dunne [45] mentions a particularly instructive dream he had in the spring of 1902, when serving in the Boer War. He seemed to be standing on a volcanic mountain. It was an island, which he had dreamed about before and knew was threatened by a catastrophic volcanic eruption (like Krakatoa). Terrified, he wanted to save the four thousand inhabitants. He tried to get the French officials on the neighbouring island to mobilize all available shipping for the rescue work. Here the dream began to develop the typical nightmare motifs of hurrying, chasing, and not arriving on time, and all the while there hovered before his mind the words: "Four thousand people will be killed unless——" A few days later Dunne received with his mail a copy of the *Daily Telegraph,* and his eye fell on the following headlines:

[45] *An Experiment with Time* (2nd edn., New York, 1938), pp. 34 *ff*.

VOLCANO DISASTER
IN MARTINIQUE

Town Swept Away

AN AVALANCHE OF FLAME

Probable Loss of Over
40,000 Lives

The dream did not take place at the moment of the actual catastrophe, but only when the paper was already on its way to him with the news. While reading it, he misread 40,000 as 4,000. The mistake became fixed as a paramnesia, so that whenever he told the dream he invariably said 4,000 instead of 40,000. Not until fifteen years later, when he copied out the article, did he discover his mistake. His unconscious knowledge had made the same mistake in reading as himself.

The fact that he dreamed this shortly before the news reached him is something that happens fairly frequently. We often dream about people from whom we receive a letter by the next post. I have ascertained on several occasions that at the moment when the dream occurred the letter was already lying in the post-office of the addressee. I can also confirm, from my own experience, the reading-mistake. During the Christmas of 1918 I was much occupied with Orphism, and in particular with the Orphic fragment in Malalas, where the Primordial Light is described as the "trinitarian Metis, Phanes, Ericepaeus." I consistently read Eric*a*paeus instead of Eric*e*paeus, as in the text. (Actually both readings occur.) This misreading became fixed as a paramnesia, and later I always remembered the name as Ericapaeus and only discovered thirty years afterward that Malalas' text has Ericepaeus. Just at this time one of my patients, whom I had not seen for a month and who knew nothing of my studies, had a dream in which an unknown man handed her a piece of paper, and on it was written a "Latin" hymn to a god called *Ericipaeus*. The dreamer was able to write this hymn down upon waking. The language it was written in was a peculiar mixture of Latin, French, and Italian. The lady had an elementary knowledge of Latin, knew a bit more Italian, and spoke French fluently. The name "Ericipaeus" was completely unknown to her, which is not surprising as she had no knowledge of the classics. Our two towns were about fifty miles apart, and there had been no communication between us for a month. Oddly enough, the variant of the name affected the very same vowel which I too had misread (*a* instead of *e*), but her unconscious misread it another way (*i* instead of *e*). I can only suppose that she unconsciously "read" not my mistake but the text in which the Latin transliteration "Ericepaeus" occurs, and was evidently put off her stroke by my misreading.

Synchronistic events rest on the *simultaneous occurrence of two different psychic states.* One of them is the normal, probable state (i.e., the one that is causally explicable), and the other, the critical experience, is the one that cannot be derived causally from the first. In the case of sudden death the critical experience cannot be recognized immediately as "extra-sensory perception" but can only be verified as such afterwards. Yet even in the case of the "scarab" what is immediately experienced is a psychic state or psychic image which differs from the dream image only because it can be verified immediately. In the case of the flock of birds there was in the woman an unconscious excitation or fear which was certainly conscious *to me* and caused me to send the patient to a heart specialist. In all these cases, whether it is a question of spatial or of temporal ESP, we find a simultaneity of the normal or ordinary state with another state or experience which is not causally derivable from it, and whose objective existence can only be verified afterwards. This definition must be borne in mind particularly when it is a question of future events. They are evidently not *synchronous* but are *synchronistic,* since they are experienced as psychic images *in the present,* as though the objective event already existed. An unexpected content which is directly or indirectly connected with some objective external event coincides with the ordinary psychic state: this is what I call synchronicity, and I maintain that we are dealing with exactly the same category of events whether their objectivity appears separated from my consciousness in space or in time. This view is confirmed by Rhine's results in so far as they were not influenced by changes in space or time. Space and time, the conceptual coordinates of bodies in motion, are probably at bottom one and the same (which is why we speak of a long or short "space of time"), and Philo Judaeus said long ago that "the extension of heavenly motion is time." [46] Synchronicity in space can equally well be conceived as perception in time,

[46] *De opificio mundi,* 26. ("Διάστημα τῆς τοῦ οὐρανοῦ κινήσεώς ἐστι ὁ χρόνος.")

but remarkably enough it is not so easy to understand synchronicity in time as spatial, for we cannot imagine any space in which future events are objectively present and could be experienced as such through a reduction of this spatial distance. But since experience has shown that under certain conditions space and time can be reduced almost to zero, causality disappears along with them, because causality is bound up with the existence of space and time and physical changes, and consists essentially in the succession of cause and effect. For this reason synchronistic phenomena cannot in principle be associated with any conceptions of causality. Hence the interconnection of meaningfully coincident factors must necessarily be thought of as acausal.

Here, for want of a demonstrable cause, we are all too likely to fall into the temptation of positing a *transcendental* one. But a "cause" can only be a demonstrable quantity. A "transcendental cause" is a contradiction in terms, because anything transcendental cannot by definition be demonstrated. If we don't want to risk the hypothesis of acausality, then the only alternative is to explain synchronistic phenomena as mere chance, which brings us into conflict with Rhine's ESP discoveries and other well-attested facts reported in the literature of parapsychology. Or else we are driven to the kind of reflections I described above, and must subject our basic principles of explanation to the criticism that space and time are constants in any given system only when they are measured without regard to psychic conditions. That is what regularly happens in scientific experiments. But when an event is observed without experimental restrictions, the observer can easily be influenced by an emotional state which alters space and time by "contraction." Every emotional state produces an alteration of consciousness which Janet called *abaissement du niveau mental;* that is to say there is a certain narrowing of consciousness and a corresponding strengthening of the unconscious which, particularly in the case of strong affects, is noticeable even to the layman. The tone of the unconscious is heightened, thereby creating a gradient for the unconscious to flow towards the conscious. The conscious then comes under the influence of unconscious instinctual impulses and contents. These are as a rule complexes whose ultimate basis is the archetype, the "instinctual pattern." The unconscious also contains subliminal perceptions (as well as forgotten memory-images that cannot be reproduced at the moment, and perhaps not at all). Among the subliminal contents we must distinguish perceptions from what I would call an inexplicable "knowledge" or "immediate existence." Whereas the perceptions can be related to possible or probable sense stimuli below the threshold of consciousness, either the "knowledge" or "immediate existence" of unconscious images has no recognizable basis, or else we find recognizable causal connections with certain already existing, and often archetypal, contents. These images, whether rooted in an already existing basis or not, stand in an analogous or equivalent (i.e., meaningful) relationship to objective occurrences which have no recognizable or even conceivable causal relationship with them. How could an event remote in space and time produce a corresponding psychic image when the transmission of energy necessary for this is not even thinkable? However incomprehensible it may appear, we are finally compelled to assume that there is in the unconscious something like an *a priori* knowledge or immediate existence of events which lacks any causal basis. At any rate our conception of causality is incapable of explaining the facts.

In view of this complicated situation it may be worth while to recapitulate the argument discussed above, and this can best be done with the aid of our examples. In Rhine's experiment I made the assumption that, owing to the tense expectation or emotional state of the subject, an already existing, correct, but unconscious image of the result enables his conscious mind to score a more than chance number of hits. The scarab dream is a conscious representation arising from an unconscious, already existing image of the situation that will occur on the following day, i.e., the recounting of the dream and the appearance of the rose-chafer. The wife of the patient who died had an unconscious knowledge of the impending death. The flock of birds evoked the corresponding memory-images and consequently her fear. Similarly, the almost simultaneous dream of the violent death of the friend arose from an already existing unconscious knowledge of it.

In all these cases and others like them there seems to be an *a priori,* causally inexplicable knowledge of a situation which is at the time unknowable. Synchronicity therefore consists of two factors: *a*) An unconscious image comes into consciousness either directly (i.e., literally) or indirectly (symbolized or suggested) in the form of a dream, idea, or premonition. *b*) An objective situation coincides with this content. The one is as puzzling as the other. How does the unconscious image arise, and how the coincidence? I understand only too well why people prefer to doubt the reality of these things. Here I will only pose the question. Later in the course of this study I will try to give an answer.

As regards the role which affects play in the occurrence of synchronistic events, I should perhaps mention that this is by no means a new idea but was already known to Avicenna and Albertus Magnus. Speaking of magic, Albertus Magnus writes:

I discovered an instructive account [of magic] in Avicenna's *Liber sextus naturalium,* which says that a certain power[47] to alter things indwells in the human soul and subordinates the other things to her, particularly when she is swept into a great excess of love or hate or the like.[48] When therefore the soul of a man falls into a great excess of any passion, it can be proved by experiment that it [the excess] binds things [magically] and alters them in the way it wants,[49] and for a long time I did not believe it, but after I had read the nigromantic books and others of the kind on signs and magic, I found that the emotionality[50] of the human soul is the chief cause of all these things, whether because, on account of her great emotion, she alters her bodily substance and the other things towards which she strives, or because, on account of her dignity, the other, lower things are subject to her, or because the appropriate hour or astrological situation or another power coincides with so inordinate an emotion, and we [in consequence] believe that what this power does is then done by the soul.[51] . . . Whoever would learn the secret of doing and undoing

---

[47] "virtus"

[48] "quando ipsa fertur in magnum amoris excessum aut odii aut alicuius talium."

[49] "fertur in grandem excessum alicuius passionis invenitur experimento manifesto quod ipse ligat res et alterat ad idem quod desiderat" [50] "affectio"

[51] "cum tali affectione exterminata concurrat hora conveniens aut ordo coelestis aut alia virtus, quae quodvis faciet, illud reputavimus tunc animam facere."

these things must know that everyone can influence everything magically if he falls into a great excess . . . and he must do it at that hour when the excess befalls him, and operate with the things which the soul prescribes. For the soul is then so desirous of the matter she would accomplish that of her own accord she seizes the more significant and better astrological hour which also rules over the things suited to that matter. . . . Thus it is the soul who desires a thing more intensely, who makes things more effective and more like what comes forth. . . . Such is the manner of production with everything the soul intensely desires. Everything she does with that aim in view possesses motive power and efficacy for what the soul desires.[52]

This text shows clearly that synchronistic ("magical") happenings are regarded as being dependent on affects. Naturally Albertus Magnus, in accordance with the spirit of his age, explains this by postulating a magical faculty in the soul, without considering that the psychic process itself is just as much "arranged" as the coinciding representation which anticipates the external physical process. This representation originates in the unconscious and therefore belongs to those "cogitationes quae sunt a nobis independentes," which, in the opinion of Arnold Geulincx, are prompted by God and do not spring from our own thinking.[53] Goethe thinks of synchronistic events in the same "magical" way. Thus he says, in his conversations with Eckermann: "We all have certain electric and magnetic powers within us and ourselves exercise an at-

[52] *De mirabilibus mundi,* incunabulum in the Zurich Zentralbibliothek, undated. (There is a Cologne printing dated 1485.)

[53] *Metaphysica vera,* Part III, "Secunda scientia," in *Opera philosophica,* ed. by J. P. N. Land, Vol. II (The Hague, 1892), pp. 187 *f.*

tractive and repelling force, according as we come into touch with something like or unlike." [54]

After these general considerations let us return to the problem of the empirical basis of synchronicity. The main difficulty here is to find an empirical material from which we can draw reasonably certain conclusions, and unfortunately this difficulty is not an easy one to solve. The experiences in question are not ready to hand. We must therefore look in the obscurest corners and summon up courage to shock the prejudices of our age if we want to broaden the basis of our understanding of nature. When Galileo discovered the moons of Jupiter with his telescope he immediately came into head-on collision with the prejudices of his learned contemporaries. Nobody knew what a telescope was and what it could do. Never before had anyone talked of the moons of Jupiter. Naturally every age thinks that all ages before it were prejudiced, and today we think this more than ever and are just as wrong as all previous ages that thought so. How often have we not seen the truth condemned! It is sad but unfortunately true that man learns nothing from history. This melancholy fact will present us with the greatest difficulties as soon as we set about collecting empirical material that would throw a little light on this dark subject, for we shall be quite certain to find it where all the authorities have assured us that nothing is to be found.

Reports of remarkable isolated cases, however well authenticated, are unprofitable and lead at most to their

[54] *Eckermann's Conversations with Goethe,* trans. by R. O. Moon (London, 1951), pp. 514 *f.* (modified).

reporter being regarded as a credulous person. Even the careful recording and verification of a large number of such cases, as in the work of Gurney, Myers, and Podmore, [55] have made next to no impression on the scientific world. The great majority of "professional" psychologists and psychiatrists seem to be completely ignorant of these researches.[56]

The results of the ESP and PK experiments have provided a statistical basis for evaluating the phenomenon of synchronicity and have at the same time pointed out the important part played by the psychic factor. This fact prompted me to ask whether it would not be possible to find a method which would on the one hand demonstrate the existence of synchronicity and, on the other hand, disclose psychic contents which would at least give us a clue to the nature of the psychic factor involved. I asked myself, in other words, whether there were not a method which would yield measurable results and at the same time give us an insight into the psychic background of synchronicity. That there are certain essential psychic conditions for synchronistic phenomena we have already seen from the ESP experiments, although the latter are in the nature of the case restricted to the fact of coincidence and

[55] Op. cit.

[56] Recently Pascual Jordan has put up an excellent case for the scientific investigation of spatial clairvoyance ("Positivistische Bemerkungen über die parapsychischen Erscheinungen," *Zentralblatt für Psychotherapie,* Leipzig, IX, 1936, no. 3). I would also draw attention to his *Verdrängung und Komplementarität* (Hamburg, 1947), concerning the relations between microphysics and the psychology of the unconscious.

only stress its psychic background without illuminating it any further. I had known for a long time that there were intuitive or "mantic" methods which start with the psychic factor and take the existence of synchronicity as self-evident. I therefore turned my attention first of all to the intuitive technique for *grasping the whole situation* which is so characteristic of China, namely the *I Ching* or *Book of Changes*.[57] Unlike the Greek-trained Western mind, the Chinese mind does not aim at grasping details for their own sake, but at a view which sees the detail as part of a whole. For obvious reasons, a cognitive operation of this kind is impossible to the unaided intellect. Judgment must therefore rely much more on the irrational functions of consciousness, that is on sensation (the "sens du réel") and intuition (perception by means of subliminal contents). The *I Ching,* which we can well call the experimental basis of classical Chinese philosophy, is one of the oldest known methods for grasping a situation as a whole and thus placing the details against a cosmic background—the interplay of Yin and Yang.

This grasping of the whole is obviously the aim of science as well, but it is a goal that necessarily lies very far off because science, whenever possible, proceeds experimentally and in all cases statistically. The experiment, however, consists in asking a definite question which excludes as far as possible anything disturbing and irrelevant. It makes conditions, imposes them on Nature, and in this way forces her to give an answer to a question devised by man. She is prevented from answering out of

[57] Trans. by Cary F. Baynes from the Richard Wilhelm translation (New York, 1950; London, 1951).

the fullness of her possibilities since these possibilities are restricted as far as practicable. For this purpose there is created in the laboratory a situation which is artificially restricted to the question and which compels Nature to give an unequivocal answer. The workings of Nature in her unrestricted wholeness are completely excluded. If we want to know what these workings are, we need a method of inquiry which imposes the fewest possible conditions, or if possible no conditions at all, and then leaves Nature to answer out of her fullness.

In the laboratory-designed experiment, the known and established procedure forms the stable factor in the statistical compilation and comparison of the results. In the intuitive or "mantic" experiment-with-the-whole, on the other hand, there is no need of any question which imposes conditions and restricts the wholeness of the natural process. It is given every possible chance to express itself. In the *I Ching* the coins fall just as happens to suit them.[58] An unknown question is followed by an unintelligible answer. Thus far the conditions for a total reaction are positively ideal. The disadvantage, however, leaps to the eye: in contrast to the scientific experiment

[58] If the experiment is made with the traditional yarrow stalks, the division of the forty-nine stalks represents the chance factor. [In his introduction to the *I Ching*, I, p. vi, Professor Jung writes: "I personified the book, in a sense, asking its judgment. . . ." On being asked by one of the editors to explain this apparent contradiction, Professor Jung replied: "When I use the *I Ching* in the case of a human individual, I ask no definite question. This is my personal choice. In China they ask specific questions. In my preface I followed this old method. Why should this be a contradiction? In my preface there is no question of a human individual."—Eds.]

one does not know what has happened. To overcome this drawback, two Chinese sages, King Wen and the Duke of Chou, in the twelfth century before our era, basing themselves on the hypothesis of the unity of nature, sought to explain the simultaneous occurrence of a psychic state with a physical process as *an equivalence of meaning*. In other words, they supposed that the same living reality was expressing itself in the psychic state as in the physical. But, in order to verify such an hypothesis, *some* limiting condition was needed in this apparently limitless experiment, namely a definite form of physical procedure, a method or technique which forced nature to answer in even and odd numbers. These, as representatives of Yin and Yang, are found both in the unconscious and in nature in the characteristic form of opposites, as the "mother" and "father" of everything that happens, and they therefore form the *tertium comparationis* between the psychic inner world and the physical outer world. Thus the two sages devised a method by which an inner state could be represented as an outer one and vice versa. This naturally presupposes an intuitive knowledge of the meaning of each oracle figure. The *I Ching*, therefore, consists of a collection of sixty-four interpretations in which the meaning of each of the possible Yin-Yang combinations is worked out. These interpretations formulate the inner unconscious knowledge that corresponds to the state of consciousness at the moment, and this psychological situation coincides with the chance results of the method, that is, with the odd and even numbers resulting from the fall of the coins or the division of the yarrow stalks.[59]

[59] See below.

The method, like all divinatory or intuitive techniques, is based on an acausal or synchronistic connective principle.[60] In practice, as any unprejudiced person will admit, many obvious cases of synchronicity occur during the experiment, which could be rationally and somewhat arbitrarily explained away as mere projections. But if one assumes that they really are what they appear to be, then they can only be meaningful coincidences for which, as far as we know, there is no causal explanation. The method consists either in dividing the forty-nine yarrow stalks into two heaps at random and counting off the heaps by threes and fives, or in throwing three coins six times, each line of the hexagram being determined by the value of obverse and reverse (heads 3, tails 2).[61] The experiment is based on a triadic principle (two trigrams) and contains sixty-four mutations, each corresponding to a psychic situation. These are discussed at length in the text and appended commentaries. There is also a Western method of very ancient origin [62] which is based on the same general principle as the *I Ching,* the only difference being that in the West this principle is not triadic but, significantly enough, tetradic, and the result is not a

[60] I first used this term in my memorial address for Richard Wilhelm (delivered May 10, 1930, in Munich). The address later appeared as an appendix to *The Secret of the Golden Flower* (London and New York, 1931), where I said: "The science of the *I Ching* is not based on the causality principle, but on a principle (hitherto unnamed because not met with among us) which I have tentatively called the *synchronistic* principle" (p. 142).

[61] *I Ching,* I, pp. 392 *f.*

[62] Mentioned by Isidore of Seville in his *Liber etymologiarum,* VIII, ix, 13 (in J. P. Migne, *Patrologia latina,* LXXXII, cols. 73–728).

hexagram built up of Yang and Yin lines but sixteen figures composed of odd and even numbers. Twelve of them are arranged according to certain rules in the astrological houses. The experiment is based on 4×4 lines consisting of a random number of points which the questioner marks in the sand or on paper from right to left.[63] In true Occidental fashion the combination of all these factors goes into considerably more detail than the *I Ching.* Here too there are any amount of meaningful coincidences, but they are as a rule harder to understand and therefore less obvious than in the latter. In the Western method, which was known since the thirteenth century as the *Ars Geomantica* or the Art of Punctation[64] and enjoyed a widespread vogue, there are no real commentaries, since its use was only mantic and never philosophical like that of the *I Ching.*

Though the results of both procedures point in the desired direction, they do not provide any basis for a statistical evaluation. I have, therefore, looked round for another intuitive technique and have hit on astrology, which, at least in its modern form, claims to give a more or less total picture of the individual's character. There is no lack of commentaries here; indeed, we find a bewildering profusion of them—a sure sign that interpretation is neither simple nor certain. The meaningful coincidence we are looking for is immediately apparent in astrology, since the astronomical data are said by astrologers to correspond to indi-

[63] Grains of corn or dice can also be used.

[64] The best account is to be found in Robert Fludd (1574–1637), *De arte geomantica.* Cf. Lynn Thorndike, *A History of Magic and Experimental Science,* II (New York, 1929), p. 110.

vidual traits of character; and from the remotest times the various planets, houses, zodiacal signs, and aspects have all had meanings that serve as a basis for a character study or for an interpretation of a given situation. Marriage can thus be "based" on ☉ ☌ ☾ in the horoscopes of the partners; or a peculiarly lucky or unlucky fate may be based on ♄ ☍, ♊ ☾, or on an unusual amount of aspects, or, again, on an old astrological maxim like *Mars in medio coeli semper significat casum ab alto,* as in the horoscope of Emperor William II. It is always possible to object that the result does not agree with our psychological knowledge of the situation or character in question, and it is difficult to refute the assertion that knowledge of character is a highly subjective affair, because in characterology there are no infallible or even reliable signs that can be in any way measured or calculated—an objection that is also raised against graphology, although in practice it enjoys widespread recognition.

This criticism, together with the absence of reliable criteria for determining traits of character, makes the meaningful coincidence of horoscope structure and individual character seem inapplicable for the purpose here under discussion. If, therefore, we want astrology to tell us anything about the acausal connection of events, we must discard this uncertain diagnosis of character and put in its place an absolutely certain and indubitable fact. Marriage is such a fact.[65]

[65] Other obvious facts would be murder and suicide. Statistics are to be found in Herbert von Kloeckler (*Astrologie als Erfahrungswissenschaft,* Leipzig, 1927, pp. 232 *ff.* and 260 *ff.*), but unfortunately they fail to give comparisons with normal average values and cannot be used for our purpose. On the other hand,

Since antiquity, the main traditional astrological and alchemical correspondence to marriage has been the *coniunctio Solis* (☉) *et Lunae* (☾), the *coniunctio Lunae et Lunae,* and the conjunction of the moon with the ascendent.[66] There are others, but these do not come within the main traditional stream. The ascendent-descendent axis was introduced into the tradition because it has long been regarded as having a particularly important influence on the

Paul Flambart (*Preuves et Bases de l'astrologie scientifique,* Paris, 1921, pp. 79 *ff.*) shows a graph of statistics on the ascendents of 123 outstandingly intelligent people. Definite accumulations occur at the corners of the airy trigon (♊, ♎, ♒). This result was confirmed by a further 300 cases.

[66] This view dates back to Ptolemy: "Apponit [Ptolemaeus] autem tres gradus concordiae: Primus cum Sol in viro, et Sol vel Luna in femina, aut Luna in utrisque, fuerint in locis se respicientibus trigono, vel hexagono aspectu. Secundus cum in viro Luna, in uxore Sol eodem modo disponuntur. Tertius si cum hoc alter alterum recipiat." (Ptolemy postulates three degrees of harmony. The first is when the sun in the man's [horoscope], and the sun or moon in the woman's, or the moon in both, are in their respective places in a trigonal or sextile aspect. The second degree is when the moon in a man's [horoscope] and the sun in a woman's are constellated in the same way. The third degree is when the one is receptive to the other.) On the same page, Cardan quotes Ptolemy (*De iudiciis astrorum*): "Omnino vero constantes et diurni convictus permanent quando in utriusque conjugis genitura luminaria contigerit configurata esse concorditer" (Generally speaking, their life together will be long and constant when in the horoscopes of both partners the luminaries [sun and moon] are harmoniously constellated). Ptolemy regards the conjunction of a masculine moon with a feminine sun as particularly favourable for marriage.—Jerome Cardan, *Commentaria in Ptolemaei librorum de iudiciis astrorum,* Book IV (in his *Opera omnia,* 1663, V, p. 332).

personality.[67] As I shall refer later to the conjunction and opposition of Mars (♂) and Venus (♀), I may say here that these are related to marriage only because the conjunction or opposition of these two planets points to a love relationship, and this may or may not produce a marriage. So far as my experiment is concerned, we have to investigate the coincident aspects ☉ ☾, ☾ ☾, and ☾ *Asc.* in the horoscopes of married pairs in relation to those of unmarried pairs. It will, further, be of interest to compare the relation of the above aspects to those of the aspects which belong only in a minor degree to the main traditional stream. No belief in astrology is needed to carry out such an investigation, only the birth-dates, an astronomical almanac, and a table of logarithms for working out the horoscope.

As the above three mantic procedures show, the method best adapted to the nature of chance is the numeri-

[67] The practising astrologer can hardly suppress a smile here, because for him these correspondences are absolutely self-evident, a classic example being Goethe's connection with Christiane Vulpius: ☉ 5° ♍ ☌ ☾ 7° ♍.

I should perhaps add a few explanatory words for those readers who do not feel at home with the ancient art and technique of astrology. Its basis is the horoscope, a circular arrangement of sun, moon, and planets according to their relative positions in the signs of the zodiac at the moment of an individual's birth. There are three main positions, viz., those of sun (☉), moon (☾), and the so-called ascendent (*Asc.*); the last has the greatest importance for the interpretation of a nativity: the *Asc.* represents the degree of the zodiacal sign rising over the eastern horizon at the moment of birth. The horoscope consists of 12 so-called "houses," sectors of 30° each. Astrological tradition ascribes different qualities to them as it does to the various "aspects," i.e., angular relations of the planets and the *luminaria* (sun ☉ and moon ☾), and to the zodiacal signs.

cal method. Since the remotest times men have used numbers to express meaningful coincidences, that is, those that can be interpreted. There is something peculiar, one might even say mysterious, about numbers. They have never been entirely robbed of their numinous aura. If, so a text-book of mathematics tells us, a group of objects is deprived of every single one of its properties or characteristics, there still remains, at the end, its *number,* which seems to indicate that number is something irreducible. (I am not concerned here with the logic of this mathematical argument, but only with its psychology!) The sequence of natural numbers turns out to be unexpectedly more than a mere stringing together of identical units: it contains the whole of mathematics and everything yet to be discovered in this field. Number, therefore, is in one sense an unpredictable entity. Although I would not care to undertake to say anything illuminating about the inner relation between two such apparently incommensurable things as number and synchronicity, I cannot refrain from pointing out that not only were they always brought into connection with one another, but that both possess numinosity and mystery as their common characteristics. Number has invariably been used to describe some numinous object, and all numbers from 1 to 9 are "sacred," just as 10, 12, 13, 14, 28, 32, and 40 have a special significance. The most elementary quality about an object is whether it is one or many. Number helps more than anything else to bring order into the chaos of appearances. It is the predestined instrument for creating order, or for apprehending an already existing, but still unknown, regular arrangement or "orderedness." It may well be the most primitive element of order in the human mind, seeing that the numbers 1 to 4 occur with the greatest frequency and have the widest incidence. In other words, primitive patterns of order are mostly triads or tetrads. That numbers have an archetypal foundation is not, by the way, a conjecture of mine but of certain mathematicians, as we shall see in due course. Hence it is not such an audacious conclusion after all if we define number psychologically as an *archetype of order* which has become conscious.[68] Remarkably enough, the psychic pictures of wholeness which are spontaneously produced by the unconscious, the symbols of the self in mandala form, also have a mathematical structure. They are as a rule quaternities (or their multiples).[69] These structures not only express order, they also create it. That is why they generally appear in times of psychic disorientation in order to compensate a chaotic state or as formulations of numinous experiences. It must be emphasized yet again that they are not inventions of the conscious mind but are spontaneous products of the unconscious, as has been sufficiently shown by experience. Naturally the conscious mind can imitate these patterns of order, but such imitations do not prove that the originals are conscious inventions. From this it follows irrefutably that the unconscious uses number as an ordering factor.

It is generally believed that numbers were *invented* or thought out by man, and are therefore nothing but concepts of quantities, containing nothing that was not previously put into them by the human intellect. But it is equally

[68] *Symbolik des Geistes* (Zurich, 1948), p. 469.

[69] Cf. *Gestaltungen des Unbewussten* (Zurich, 1950), pp. 95 *ff.* and 189 *ff.*

possible that numbers were *found* or discovered. In that case they are not only concepts but something more—autonomous entities which somehow contain more than just quantities. Unlike concepts, they are based not on a psychic assumption but on the quality of being themselves, on a "so-ness" that cannot be expressed by an intellectual concept. Under these conditions they might easily be endowed with qualities that have still to be discovered. I must confess that I incline to the view that numbers were as much found as invented, and that in consequence they possess a relative autonomy analogous to that of the archetypes. They would then have, in common with the latter, the quality of being pre-existent to consciousness, and hence, on occasion, of conditioning it rather than being conditioned by it. The archetypes too, as *a priori* ideal forms, are as much found as invented: they are *discovered* inasmuch as one did not know about their unconscious autonomous existence, and *invented* inasmuch as their presence was inferred from analogous conceptual structures. Accordingly it would seem that natural numbers have an archetypal character. If that is so, then not only would certain numbers and combinations of numbers have a relation to and an effect on certain archetypes, but the reverse would also be true. The first case is equivalent to number magic, but the second is equivalent to inquiring whether numbers, in conjunction with the combination of archetypes found in astrology, would show a tendency to behave in a special way.

CHAPTER TWO

AN ASTROLOGICAL EXPERIMENT

As I have already said, we need two different facts, one of which represents the astrological constellation, and the other the married state.

The material to be examined, namely a quantity of marriage horoscopes, was obtained from friendly donors in Zurich, London, Rome, and Vienna. Originally the material had been put together for purely astrological purposes, some of it many years ago, so that those who gathered the material knew of no connection between its collection and the aim of the present study, a fact which I stress because it might be objected that the material was specially selected with that aim in view. This was not so; the sample was a random one. The horoscopes, or rather the birth data, were piled up in chronological order just as the post brought them in. When the horoscopes of 180 married pairs had come in, there was a pause in the collection, during which the 360 horoscopes were worked out. This first batch was used to conduct a pilot investigation, as I wanted to test out the methods to be employed.

Since the material had been collected originally in order to test the empirical foundations of this intuitive method, a few more general remarks may not be out of place concerning the considerations which prompted the collection of the material.

Marriage is a clear-cut fact, although its psychological content shows every conceivable variation. According to the astrological view, it is precisely this fact of marriage that expresses itself markedly in the horoscopes. The possibility that the individuals characterized by the horoscopes married one another, so to say, by accident will necessarily recede into the background; all external factors seem capable of astrological evaluation, but only inasmuch as they are represented psychologically. Owing to the very large number of characterological variations, we would hardly expect marriage to be characterized by only *one* astrological configuration; rather, if astrological assumptions are at all correct, there will be several configurations that point to a predisposition in the choice of a marriage partner. In this connection I must call the reader's attention to the well-known correspondence between the sun-spot periods and the mortality curve. The connecting link appears to be the disturbances of the earth's magnetic field, which in their turn are due to fluctuations in the proton radiation from the sun. These fluctuations also have an influence on "radio weather" by disturbing the ionosphere that reflects the radio waves.[1] Investigation of these disturbances seems to

[1] For a comprehensive account of this, see Max Knoll, "Transformations of Science in Our Time," in *Man and Time* (Papers from the Eranos Yearbooks, 3; New York and London, in press; orig. in *Eranos-Jahrbuch 1951*).

indicate that the conjunctions, oppositions, and quadratic aspects of the planets play a considerable part in increasing the proton radiation and thus causing electromagnetic storms. On the other hand, the astrologically favourable trigonal and sextile aspects have been reported to produce uniform radio weather.

These observations give us an unexpected glimpse into a possible causal basis for astrology. At all events, this is certainly true of Kepler's weather astrology. But it is also possible that, over and above the already established physiological effects of proton radiation, psychic effects can occur which would rob astrological statements of their chance nature and bring them within range of a causal explanation. Although nobody knows what the validity of a nativity horoscope rests on, it is just conceivable that there is a causal connection between the planetary aspects and the psycho-physiological disposition. One would therefore do well not to regard the results of astrological observation as synchronistic phenomena, but to take them as possibly causal in origin. For, wherever a cause is even remotely thinkable, synchronicity becomes an exceedingly doubtful proposition.

For the present, at any rate, we have insufficient grounds for believing that the astrological results are more than mere chance, or that statistics involving large numbers yield a statistically significant result.[2] As large-scale studies are lacking, I decided to investigate the empirical basis of astrology, using a large number of horoscopes of married pairs just to see what kind of figures would turn up.

[2] Cf. the statistical results in K. E. Krafft, *Traité d'astro-biologie* (Paris, 1939), pp. 23 *ff.* and passim.

Pilot Investigation

With the first batch assembled, I turned first to the conjunctions (☌) and oppositions (☍) of sun and moon,[3] two aspects regarded in astrology as being about equally strong (though in opposite senses), i.e., as signifying intensive relations between the heavenly bodies. With the ♂, ♀, *Asc.*, and *Desc.* conjunctions and oppositions, they yield fifty different aspects.[4]

Male

		☉	☾	♂	♀	*Asc.*	*Desc.*
	☉	☌ ☍	☌ ☍	☌ ☍	☍ ☌	☌	☌
	☾	☌ ☍	☌ ☍	☍ ☌	☌ ☍	☌	☌
Female	♂	☍ ☌	☌ ☍	☌ ☍	☌ ☍	☌	☌
	♀	☌ ☍	☍ ☌	☌ ☍	☍ ☌	☌	☌
	Asc.	☌	☌	☌	☌	☌	☌
	Desc.	☌	☌	☌	☌		

☌ = conjunction ☍ = opposition

FIG. 1

[3] Although the quadratic, trigonal, and sextile aspects and the relations to the Medium and Imum Coeli ought really to be considered, I have omitted them here so as not to make the exposition unduly complicated. The main point is not what marriage aspects are, but whether they can be detected in the horoscope.

[4] The chart in Fig. 1 sets out clearly the 50 different aspects as they actually occurred in the 180 married pairs.

The reasons why I chose these combinations will be clear to the reader from my remarks on the astrological traditions in the previous chapter. I have only to add here that, of the conjunctions and oppositions, those of Mars and Venus are far less important than the rest, as will readily be appreciated from the following consideration: the relation of Mars to Venus can reveal a love relation, but a marriage is not always a love relation and a love relation is not always a marriage. My aim in including the conjunction and opposition of Mars and Venus was therefore to compare them with the other conjunctions and oppositions.

These fifty aspects were first studied for 180 married couples. It is clear that these 180 men and 180 women can also be paired off into unmarried couples. In fact, since any one of the 180 men could be paired off with any one of the 179 women to whom he was not married, it is clear that we can investigate $180 \times 179 = 32{,}220$ unmarried pairs within the group of 180 marriages. This was done (cf. Table I), and the aspect analysis for these unmarried pairs was compared with that for the married pairs. For all calculations, an orbit of 8° either way was assumed, clockwise and anticlockwise, not only inside the sign but extending beyond it. Later, additional marriages were added to the original batch, so that, in all, 483 marriages, or 966 horoscopes, were examined. As the following account shows, the testing and the tabulation of results were carried out in batches.

To begin with, what interested me most was, of course, the question of probability: were the maximum results that we obtained "significant" figures or not?—that is, were they improbable or not? Calculations undertaken by a mathematician showed unmistakably that the average frequency of 10% in the first batch and subsequently in all three batches is far from representing a significant figure. Its probability is much too great; in other words, there is no ground for assuming that our maximum frequencies are more than mere dispersions due to chance.

Analysis of First Batch

First we counted all the conjunctions and oppositions between ☉ ☾ ♂ ♀ *Asc.* and *Desc.* for the 180 married and the 32,220 unmarried pairs. The results are shown in Table I, where it will be observed that the aspects are arranged by frequency of their occurrence in the married and unmarried pairs.

Clearly, the frequencies of occurrence shown in columns 2 and 4 of Table I for observed occurrences of the aspects in married and unmarried pairs respectively are not immediately comparable, since the first are occurrences in 180 pairs and the second in 32,220 pairs.[5] In column 5, therefore, we show the figures in column 4 multiplied by the factor $\frac{180}{32{,}220}$. If the right side (unmarried pairs) = 1, then we get the following proportion: 18 : 8.40 = 2.14 : 1. In

[5] [In this way a rough control group is obtained. It will, however, be appreciated that it is derived from a much larger number of pairs than the married pairs: 32,220 as compared with 180. This leads to the possibility of showing the chance nature of the 180 pairs. On the hypothesis that all the figures are due to chance, we would expect a far greater accuracy in the greater number and consequently a much smaller range in the figures. This is so, for the range in the 180 married pairs is 18 — 2 = 16, whereas in the 180 unmarried pairs we get 9.6 — 7.4 = 2.2.—EDS.]

TABLE I

Fem.	Aspect	Masc.	Observed Occurrences for 180 Married Pairs: Actual Occurrences	Observed Occurrences for 180 Married Pairs: Percentage Occurrences	Observed Occurrences for 32,220 Unmarried Pairs	Calculated Frequency for 180 Unmarried Pairs: Actual Frequency	Calculated Frequency for 180 Unmarried Pairs: Frequency Percentage
Moon	☌	Sun	18	10.0%	1506	8.4	4.7
Asc.	☌	Venus	15	8.3%	1411	7.9	4.4
Moon	☌	Asc.	14	7.7%	1485	8.3	4.6
Moon	☍	Sun	13	7.2%	1438	8.0	4.4
Moon	☌	Moon	13	7.2%	1479	8.3	4.6
Venus	☍	Moon	13	7.2%	1526	8.5	4.7
Mars	☌	Moon	13	7.2%	1548	8.6	4.8
Mars	☌	Mars	13	7.2%	1711	9.6	5.3
Mars	☌	Asc.	12	6.6%	1467	8.2	4.6
Sun	☌	Mars	12	6.6%	1485	8.3	4.6
Venus	☌	Asc.	11	6.1%	1409	7.9	4.4
Sun	☌	Asc.	11	6.1%	1413	7.9	4.4
Mars	☌	Desc.	11	6.1%	1471	8.2	4.6
Desc.	☌	Venus	11	6.1%	1470	8.2	4.6
Venus	☌	Desc.	11	6.1%	1526	8.5	4.7
Moon	☍	Mars	10	5.5%	1540	8.6	4.8
Venus	☍	Venus	9	5.0%	1415	7.9	4.4
Venus	☌	Mars	9	5.0%	1498	8.4	4.7
Venus	☌	Sun	9	5.0%	1526	8.5	4.7
Moon	☌	Mars	9	5.0%	1539	8.6	4.8
Sun	☌	Desc.	9	5.0%	1556	8.7	4.8
Asc.	☌	Asc.	9	5.0%	1595	8.9	4.9
Desc.	☌	Sun	8	4.3%	1398	7.8	4.3
Venus	☍	Sun	8	4.3%	1485	8.3	4.6
Sun	☌	Moon	8	4.3%	1508	8.4	4.7

TABLE I (*continued*)

Fem.	Aspect	Masc.	Observed Occurrences for 180 Married Pairs		Observed Occurrences for 32,220 Unmarried Pairs	Calculated Frequency for 180 Unmarried Pairs	
			Actual Occurrences	Percentage Occurrences		Actual Frequency	Frequency Percentage
Sun	☍	Venus	8	4.3%	1502	8.4	4.7
Sun	☍	Mars	8	4.3%	1516	8.5	4.7
Mars	☍	Sun	8	4.3%	1516	8.5	4.7
Mars	☌	Venus	8	4.3%	1520	8.5	4.7
Venus	☍	Mars	8	4.3%	1531	8.6	4.8
Asc.	☌	Moon	8	4.3%	1541	8.6	4.8
Moon	☍	Moon	8	4.3%	1548	8.6	4.8
Desc.	☌	Moon	8	4.3%	1543	8.6	4.8
Asc.	☌	Mars	8	4.3%	1625	9.1	5.0
Moon	☌	Venus	7	3.8%	1481	8.3	4.6
Mars	☍	Venus	7	3.8%	1521	8.5	4.7
Moon	☌	Desc.	7	3.8%	1539	8.6	4.8
Mars	☍	Moon	7	3.8%	1540	8.6	4.8
Asc.	☌	Desc.	6	3.3%	1328	7.4	4.1
Desc.	☌	Mars	6	3.3%	1433	8.0	4.4
Venus	☌	Moon	6	3.3%	1436	8.0	4.4
Asc.	☌	Sun	6	3.3%	1587	8.9	4.9
Mars	☌	Sun	6	3.3%	1575	8.8	4.9
Moon	☍	Venus	6	3.3%	1576	8.8	4.9
Venus	☌	Venus	5	2.7%	1497	8.4	4.7
Sun	☍	Moon	5	2.7%	1530	8.6	4.8
Sun	☌	Venus	4	2.2%	1490	8.3	4.6
Mars	☍	Mars	3	1.6%	1440	8.0	4.4
Sun	☌	Sun	2	1.1%	1480	8.3	4.6
Sun	☍	Sun	2	1.1%	1482	8.3	4.6

TABLE II

Aspect Fem.		Aspect Masc.	Proportion of Aspect Frequencies for Married Pairs
Moon	☌	Sun	2.14
Asc.	☌	Venus	1.89
Moon	☌	Asc.	1.68
Moon	☍	Sun	1.61
Moon	☌	Moon	1.57
Venus	☍	Moon	1.53
Mars	☌	Moon	1.50
Mars	☌	Asc.	1.46
Sun	☌	Mars	1.44
Venus	☌	Asc.	1.39
Sun	☌	Asc.	1.39
Mars	☌	Mars	1.36
Mars	☌	Desc.	1.34
Desc.	☌	Venus	1.34
Venus	☌	Desc.	1.29
Moon	☍	Mars	1.16
Venus	☍	Venus	1.14
Venus	☌	Mars	1.07
Venus	☌	Sun	1.06
Moon	☌	Mars	1.05
Sun	☌	Desc.	1.04
Desc.	☌	Sun	1.02
Asc.	☌	Asc.	1.01
Venus	☍	Sun	0.96
Sun	☌	Moon	0.95

Table II, these proportions are arranged according to frequency.

To a statistician, these numbers cannot be used to confirm anything, and so are valueless, because they are chance dispersions. But on psychological grounds I have discarded the idea that we are dealing with *mere* chance numbers. In

TABLE II (*continued*)

Aspect Fem.		Masc.	Proportion of Aspect Frequencies for Married Pairs
Sun	☍	Venus	0.95
Sun	☍	Mars	0.94
Mars	☍	Sun	0.94
Mars	☌	Venus	0.94
Venus	☍	Mars	0.94
Asc.	☌	Moon	0.93
Moon	☍	Moon	0.93
Desc.	☌	Moon	0.92
Asc.	☌	Mars	0.88
Moon	☌	Venus	0.85
Mars	☍	Venus	0.82
Moon	☌	Desc.	0.81
Asc.	☌	Desc.	0.81
Mars	☍	Moon	0.81
Desc.	☌	Mars	0.75
Venus	☌	Moon	0.75
Asc.	☌	Sun	0.68
Mars	☌	Sun	0.68
Moon	☍	Venus	0.68
Venus	☌	Venus	0.60
Sun	☍	Moon	0.59
Sun	☌	Venus	0.48
Mars	☍	Mars	0.37
Sun	☌	Sun	0.24
Sun	☍	Sun	0.24

a total picture of natural events, it is just as important to consider the exceptions to the rule as the averages. This is the fallacy of the statistical picture: it is one sided, inasmuch as it represents only the average aspect of reality and excludes the total picture. The statistical view of the world is a mere abstraction and therefore incomplete and even

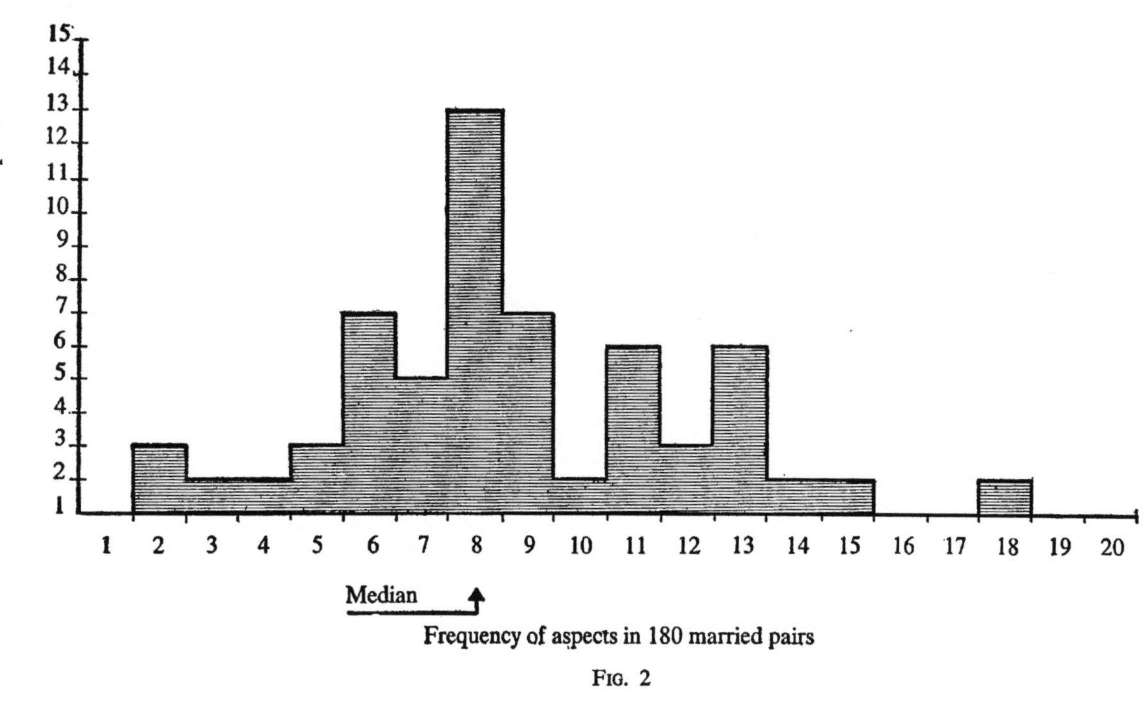
Aspects occurring a stated number of times in 180 married pairs
15
14
13
12
11
10
9
8
7
6
5
4
3
2
1
1 2 3 4 5 6 7 8 9 10 11 12 13 14 15 16 17 18 19 20
Median
Frequency of aspects in 180 married pairs

FIG. 2

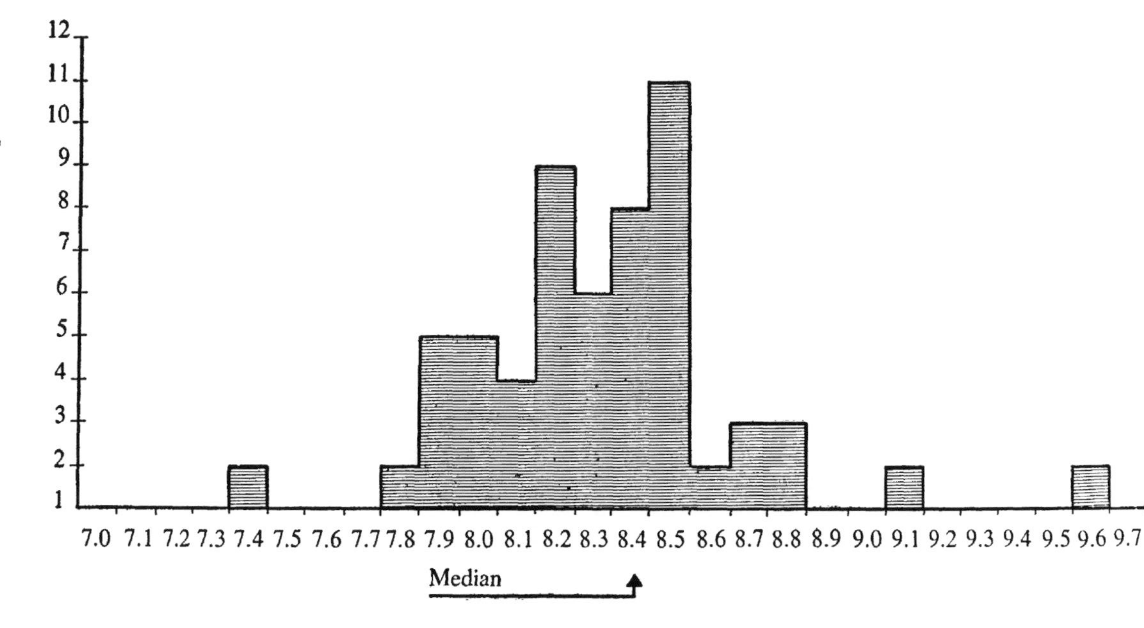

Frequency of aspects per 180 unmarried pairs, observed in 32,220 pairs

FIG. 3

fallacious, particularly so when it deals with man's psychology. Inasmuch as chance maxima and minima occur, they are *facts* whose nature I set out to explore.

What strikes us in Table II is the unequal distribution of the frequency values. The top seven and bottom six aspects both show a fairly strong dispersion, while the middle values tend to cluster round the proportion 1 : 1. I shall come back to this peculiar distribution with the help of a special graph (Fig. 2).

An interesting point is the confirmation of the traditional astrological and alchemical correspondence between marriage and the moon-sun aspects:

(fem.) moon ☌ (masc.) sun 2.14 : 1
(fem.) moon ☍ (masc.) sun 1.61 : 1

whereas there is no evidence of any emphasis on the Venus-Mars aspects.

Of the fifty possible aspects, the result shows that for the married pairs there are fifteen such configurations whose frequency is well above the proportion 1 : 1. The highest value is found in the aforementioned moon-sun conjunction, and the two next-highest figures—1.89 : 1 and 1.68 : 1—correspond to the conjunctions between (fem.) *Asc.* and (masc.) Venus, or (fem.) moon and (masc.) *Asc.*, thus apparently confirming the traditional significance of the ascendent.

Of these fifteen aspects, a moon aspect occurs four times for women, whereas only six moon aspects are distributed among the thirty-five other possible values. The mean proportional value of all moon aspects amounts to 1.24 : 1. The average value of the four just cited in the table amounts to 1.74 : 1, as compared with 1.24 : 1 for all moon aspects. The moon seems to be less emphasized for men than for women.

For men the corresponding role is played not by the sun but by the *Asc.-Desc.* axis. In the first fifteen aspects of Table II, these aspects occur six times for men and only twice for women. In the former case they have an average value of 1.42 : 1, as compared with 1.22 : 1 for all masculine aspects between *Asc.* or *Desc.* on the one hand and one of the four heavenly bodies on the other.

Figures 2 and 3 give a graphic representation of the values listed in Figure 1 from the point of view of the dispersion of aspects.

This arrangement enables us not only to visualize the dispersions in the frequency of occurrence of the different aspects but also to make a rapid estimate of the mean number of occurrences per aspect, using the median as an estimator. Whereas, in order to get the arithmetic mean, we have to total the aspect frequencies and divide by the number of aspects, the median frequency is found by counting down the histogram to a point where half the squares are counted and half are still to count. Since there are fifty squares in this case, the median is seen to be 8.0, since 25 squares do not exceed this value and 25 squares do exceed it (cf. Fig. 2).

For the married pairs the median amounts to 8 cases, but in the combinations of unmarried persons it is more, namely 8.4 (cf. Fig. 3). For the unmarried the median coincides with the arithmetical mean—both amount to 8.4 —whereas the median for the married is lower than the corresponding mean value of 8.4, which is due to the pres-

TABLE III

First Batch				Second Batch				Both Batches			
180 Married Pairs				220 Married Pairs				400 Married Pairs			
Moon	☌	Sun	10.0%	Moon	☌	Moon	10.9%	Moon	☌	Moon	9.2%
Asc.	☌	Venus	9.4%	Mars	☍	Venus	7.7%	Moon	☍	Sun	7.0%
Moon	☌	Asc.	7.7%	Venus	☌	Moon	7.2%	Moon	☌	Sun	7.0%
Moon	☌	Moon	7.2%	Moon	☍	Sun	6.8%	Mars	☌	Mars	6.2%
Moon	☍	Sun	7.2%	Moon	☍	Mars	6.8%	Desc.	☌	Venus	6.2%
Mars	☌	Moon	7.2%	Desc.	☌	Mars	6.8%	Moon	☍	Mars	6.2%
Venus	☍	Moon	7:2%	Desc.	☌	Venus	6.3%	Mars	☌	Moon	6.0%
Mars	☌	Mars	7.2%	Moon	☍	Venus	6.3%	Mars	☍	Venus	5.7%
Mars	☌	Asc.	6.6%	Venus	☌	Venus	6.3%	Moon	☌	Asc.	5.7%
Sun	☌	Mars	6.6%	Sun	☍	Mars	5.9%	Venus	☌	Desc.	5.7%
Venus	☌	Desc.	6.1%	Venus	☌	Desc.	5.4%	Venus	☌	Moon	5.5%
Venus	☌	Asc.	6.1%	Venus	☌	Mars	5.4%	Desc.	☌	Mars	5.2%
Mars	☌	Desc.	6.1%	Sun	☌	Moon	5.4%	Asc.	☌	Venus	5.2%
Sun	☌	Asc.	6.1%	Sun	☌	Sun	5.4%	Sun	☍	Mars	5.2%

ence of lower values for the married pairs. A glance at Figure 2 will show that there is a wide dispersion of values which contrasts strikingly with those clustered round the mean figure of 8.4 in Figure 3. Here there is not a single aspect with a frequency greater than 9.6 (cf. Fig. 3), whereas among the married one aspect reaches a frequency of nearly twice as much (cf. Fig. 2).

Comparison of All Batches

On the supposition that the dispersion apparent in Figure 2 was due to chance I investigated a larger number of marriage horoscopes, four hundred in all (or eight hundred individual horoscopes). The results of this additional material are shown in Table III and set against the 180 cases already discussed, though I have here confined myself to the maximal numbers that clearly exceed the median. Figures are given in percentages.

The 180 married couples in the first column show the results of the first collection, while the 220 in the second column were collected more than a year later. The second column not only differs from the first in its aspects, but shows a marked sinking of the frequency values. The only exception is the top number, representing the classical ☾ ☌ ☾. It takes the place of the equally classical ☾ ☌ ☉ in the first column. Of the fourteen aspects in the first column only four come up again in the second, but of these no less than three are moon aspects, and this is in accord with astrological expectations. The absence of correspondence between the aspects of the first and second columns indicates a great inequality of material, i.e., there is a wide dispersion. One can see this in the aggregate

TABLE IV

Frequency in %	☾ ☌ ☉	☾ ☌ ☾	☾ ☍ ☉	Average
180 Married Pairs	10.0	7.2	7.2	8.1
220 Married Pairs	4.5	10.9	6.8	7.4
180 + 220 = 400 Married Pairs	7.0	9.2	7.0	7.7
83 Additional Married Pairs	7.2	4.8	4.8	5.6
83 + 400 = 483 Married Pairs	7.2	8.4	6.6	7.4

figures for the 400 married pairs: as a result of the evening out of the dispersion they all show a marked decrease. These proportions are brought out still more clearly in Table IV.

This table shows the frequency figures for the three constellations that occur most often: two lunar conjunctions and one lunar opposition. The highest average frequency, that for the original 180 marriages, is 8.1%; for the 220 collected and worked out later the average maximum drops to 7.4%; and for the 83 marriages that were added still later the average amounts to only 5.6%. In the original batches of 180 and 220 the maxima still lie with the same aspects, but in the last batch of 83 the maxima lie with different aspects, namely *Asc.* ☌ ☾, ☉ ☌ ♀, ☉ ☌ ♂, and *Asc.* ☌ *Asc.* The average maximum for these four aspects is 8.7%. This high figure exceeds our highest average of 8.1% for the first batch of 180, which only proves how fortuitous our "favourable" initial results were. Nevertheless it is worth pointing out that, amusingly enough, the maximum of 9.6% [6] lies with

[6] [I.e., 8 in 83.—EDS.]

TABLE V

Maximal Frequency in % for

1. 300 pairs combined at random	7.3
2. 325 pairs chosen by lot	6.5
3. 400 pairs chosen by lot	6.2
4. 32,220 pairs	5.3

the *Asc.* ☌ ☾ aspect, that is to say, with another lunar aspect which is supposed to be particularly characteristic of marriage—a *lusus naturae,* no doubt, but a very queer one, since according to tradition the ascendent or "horoscopus," together with sun and moon, forms the trinity that determines fate and character. Had one wanted to falsify the statistical findings so as to bring them into line with tradition one could not have done it more successfully.

Table V gives the maximal frequencies for unmarried pairs. The first result was obtained by my co-worker, Dr. Liliane Frey-Rohn, putting the men's horoscopes on one side and the women's on the other, and then combining each of the pairs that happened to lie on top. Care was naturally taken that a real married pair was not accidentally combined. The resultant frequency of 7.3 is pretty high in comparison with the much more probable maximal figure for the 32,220 unmarried pairs, which is only 5.3. This first result seemed to me somewhat suspicious.[7] I therefore suggested that we should not combine

[7] How subtle these things can be is shown by the following incident: Recently it fell to my colleague to make the table arrangement for a number of people who were invited to dinner. She did this with care and discretion. But at the last moment an esteemed guest, a man, unexpectedly turned up who had at all

the pairs ourselves, but should proceed in the following way: 325 men's horoscopes were numbered, the numbers were written on separate slips, thrown into a pot, and mixed up. Then a person who knew nothing of astrology and psychology and even less of these investigations was invited to draw the slips one by one out of the pot, without looking at them. The numbers were each combined with the topmost on the pile of women's horoscopes, care being again taken that married pairs did not accidentally come together. In this way 325 artificial pairs were obtained. The resultant 6.5 is rather nearer to probability. Still more probable is the result obtained for the 400 unmarried pairs. Even so, this figure (6.2) is still too high.

The somewhat curious behaviour of our figures led to a further experiment whose results I mention here with

costs to be suitably placed. The table arrangement was all upset, and a new one had to be hastily devised. There was no time for elaborate reflection. As we sat down to table, the following astrological picture manifested itself in the immediate vicinity of the guest:

LADY ☾ in ♌	LADY ☉ in ♓	GUEST ☉ in ♉	LADY ☉ in ♓
LADY ☉ in ♌	LADY ☾ in ♓	GENTLEMAN ☾ in ♉	LADY ☾ in ♓

Four ☉ ☾ marriages had arisen. My colleague, of course, had a thorough knowledge of astrological marriage aspects, and she was also acquainted with the horoscopes of the people in question. But the speed with which the new table arrangement had to be made left her no opportunity for reflection, so that the unconscious had a free hand in secretly arranging the "marriages."

all the necessary reserve, though it seems to me to throw some light on the statistical variations. It was made with three people whose psychological status was accurately known. The experiment consisted in taking 400 marriage horoscopes at random and providing 200 of them with numbers. Twenty of these were then drawn by lot by the subject. These twenty married pairs were examined statistically for our fifty marriage characteristics. The first subject was a woman patient who, at the time of the experiment, found herself in a state of intense emotional excitement. It proved that of twenty Mars aspects no less than ten were emphasized, with a frequency of 15.0; of the moon aspects nine, with a frequency of 10.0; and of the sun aspects nine, with a frequency of 14.0. The classical significance of Mars lies in his emotionality, in this case supported by the masculine sun. As compared with our general results there is a predominance of the Mars aspects, which fully agrees with the psychic state of the subject.

The second subject was a woman patient whose main problem was to realize and assert her personality in the face of her self-suppressive tendencies. In this case the axial aspects (*Asc. Desc.*), which are supposed to be characteristic of the personality, came up twelve times with a frequency of 20.0, and the moon aspects with a frequency of 18.0. This result, astrologically considered, was in full agreement with the subject's actual problems.

The third subject was a woman with strong inner oppositions whose union and reconciliation constituted her main problem. The moon aspects came up fourteen times with a frequency of 20.0, the sun aspects twelve times with a frequency of 15.0, and the axial aspects nine times with a frequency of 14.0. The classical *coniunctio Solis et Lunae* as the symbol of the union of opposites is clearly emphasized.

In all these cases the selection by lot of marriage horoscopes proves to have been influenced, and this fits in with our experience of the *I Ching* and other mantic procedures. Although all these figures lie well within the limits of probability and cannot therefore be regarded as anything more than chance, their variation, which each time corresponds surprisingly well with the psychic state of the subject, still gives one food for thought. The psychic state was characterized as a situation in which insight and decision come up against the insurmountable barrier of an unconscious opposed to the will. This relative defeat of the powers of the conscious mind constellates the moderating archetype, which appears in the first case as Mars, the emotional *maleficus,* in the second case as the equilibrating axial system that strengthens the personality, and in the third case as the *hieros gamos* or *coniunctio* of supreme opposites.[8] The psychic and the physical event (namely, the subject's problems and choice of horoscope) correspond, it would seem, to the nature of the archetype in the background and could therefore represent a synchronistic phenomenon.

Inasmuch as I am not very well up in the higher mathematics, and had therefore to rely on the help of a profes-

[8] Cf. the nuptials of sun and moon in alchemy: *Psychology and Alchemy* (New York and London, 1953), index, s.v. "sun and moon."

sional, I asked Professor Markus Fierz, of Basel, to calculate the probability of my maximal numbers.[9] This he very kindly did, and using the Poisson distribution he arrived at a probability of approximately 1 : 10,000. Later, on checking the calculation, he found an error whose correction raised the probability to 1 : 1500.[10] From this it is clear that although our best results—☾ ☌ ☉ and ☾ ☌ ☾—are fairly improbable in practice, they are theoretically so probable that there is little justification for regarding the immediate results of our statistics as anything more than chance. If for instance there is a 1 : 1500 probability of my getting the telephone connection I want, I shall probably prefer, instead of waiting on the off-chance for a telephone conversation, to write a letter. Our investigation shows that not only do the frequency values approximate to the average with the greatest number of married pairs, but that any chance pairings produce similar statistical proportions. From the scientific point of view the result of our investigation is in some respects not encouraging for astrology, as everything seems to indicate that in the case of large numbers the differences between the frequency values for the marriage aspects of married and unmarried persons disappear altogether. Thus, from the scientific point of view, there is little hope of proving that

[9] [See the appendix to this chapter.—EDS.]

[10] Professor Fierz wishes to correct this sentence as follows: "Later on he called my attention to the fact that the sequence of the 3 aspects does not matter. As there are 6 possible sequences, we have to multiply our probability by 6, which gives 1 : 1500." To this I reply that I never suggested anything of the kind! The sequence, i.e., the way in which the 3 conjunctions follow each other, has no importance at all.

astrological correspondence is something that conforms to law. At the same time, it is not so easy to counter the astrologer's objection that my statistical method is too arbitrary and too clumsy to evaluate correctly the numerous psychological and astrological aspects of marriage.

So the essential thing that remains over from our astrological statistics is the fact that the first batch of 180 marriage horoscopes shows a distinct maximum of 18 for ☾ ☌ ☉ and the second batch of 220 a maximum of 24 for ☾ ☌ ☾. These two aspects have long been mentioned in the old literature as marriage characteristics, and they therefore represent the oldest tradition. The third batch of 83 yields, as we have said, a maximum of 8 for ☾ ☌ *Asc.* These batches have probabilities of about 1 : 1000, 1 : 10,000, and 1 : 50 respectively. I should like to illustrate what has happened here by means of an example:

You take three matchboxes, put 1,000 black ants in the first, 10,000 in the second and 50 in the third, together with one white ant in each, shut the boxes, and bore a hole in each of them, small enough to allow only one ant to crawl through at a time. The first ant to come out of each of the three boxes is always the white one.

The chances of this actually happening are extremely improbable. Even in the first two cases, the probability works out at 1 : 1000 × 10,000, which means that such a coincidence is to be expected only in one case out of 10,000,000. It is improbable that it would ever happen in anyone's experience. Yet in my statistical investigation it happened that precisely the three conjunctions stressed by astrological tradition came together in the most improbable way.

For the sake of accuracy, however, it should be pointed out that it is not the *same* white ant that is the first to appear each time. That is to say, although there is always a lunar conjunction and always a "classical" one of decisive significance, they are nevertheless different conjunctions, because each time the moon is associated with a different partner. These are of course the three main components of the horoscope, namely the ascendent, or rising degree of a zodiacal sign, which characterizes the moment; the moon, which characterizes the day; and the sun, which characterizes the month of birth. Hence, if we consider only the first two batches, we must assume two white ants for each box. This correction raises the probability of the coinciding lunar conjunctions to 1 : 2,500,000. If we take the third batch as well, the coincidence of the three classical moon aspects has a probability of 1 : 62,500,000. The first proportion is significant even when taken by itself, for it shows that the coincidence is a very improbable one. But the coincidence with the third lunar conjunction is so remarkable that it looks like a deliberate arrangement in favour of astrology. If, therefore, the result of our experiment should be found to have a significant—i.e., more than merely chance—probability, the case for astrology would be proved in the most satisfactory way. If, on the contrary, the figures actually fall within the limits of chance probability, they do not support the astrological claim, they merely *imitate* accidentally the ideal answer to astrological expectation. It is nothing but a chance result from the statistical point of view, yet it is *meaningful* on account of the fact that it looks as if it validated this expectation. It is just what I call a synchronistic phenomenon. The statistically significant statement only concerns regularly occurring events, and if considered as axiomatic, it simply abolishes all exceptions to the rule. It produces a merely average picture of natural events, but not a *true* picture of the world as it is. Yet the exceptions—and my results are exceptions and most improbable ones at that—are just as important as the rules. Statistics would not even make sense without the exceptions. There is no rule that is true under all circumstances, for this is the real and not a statistical world. Because the statistical method shows only the average aspects, it creates an artificial and predominantly conceptual picture of reality. That is why we need a complementary principle for a complete description and explanation of nature.

If we now consider the results of Rhine's experiments, and particularly the fact that they depend in large measure on the subject's active interest,[11] we can regard what happened in our case as a synchronistic phenomenon. The statistical material shows that a practically as well as theoretically improbable chance combination occurred which coincides in the most remarkable way with traditional astrological expectations. That such a coincidence should occur at all is so improbable and so incredible that nobody could have dared to predict anything like it. It

[11] Cf. G. Schmeidler, "Personality Correlates of ESP as Shown by Rorschach Studies," *Journal of Parapsychology* (Durham), XIII (1950), 23 *ff.* The author points out that those who accept the possibility of ESP get results above expectation, whereas those who reject it get negative results.

really does look as if the statistical material had been manipulated and arranged so as to give the appearance of a positive result. The necessary emotional and archetypal conditions for a synchronistic phenomenon were already given, since it is obvious that both my co-worker and myself had a lively interest in the outcome of the experiment, and apart from that the question of synchronicity had been engaging my attention for many years. What seems in fact to have happened—and seems often to have happened, bearing in mind the long astrological tradition—is that we got a result which has presumably turned up many times before in history. Had the astrologers (with but few exceptions) concerned themselves more with statistics and questioned the justice of their interpretations in a scientific spirit, they would have discovered long ago that their statements rested on a precarious foundation. But I imagine that in their case too, as with me, a secret, mutual connivance existed between the material and the psychic state of the astrologer. This correspondence is simply *there* like any other agreeable or annoying accident, and it seems doubtful to me whether it can be proved scientifically to be anything more than that.[12] One may be fooled by coincidence, but one has to have a very thick skin not to be impressed by the fact that, out of fifty possibilities, three times precisely those turned up as maxima which are regarded by tradition as typical.

[12] As my statistics show, the result becomes blurred with larger figures. So it is very probable that if more material were collected it would no longer produce a similar result. We have therefore to be content with this apparently unique *lusus naturae,* though its uniqueness in no way prejudices the facts.

As though to make this startling result even more impressive, we found that use had been made of unconscious deception. On first working out the statistics I was put off the trail by a number of errors which I fortunately discovered in time. After overcoming this difficulty I then forgot to mention, in the Swiss edition of this book, that the ant comparison, if applied to our experiment, only fits if respectively two or three white ants are assumed each time. This considerably reduces the improbability of our results. Then, at the eleventh hour, Professor Fierz, on checking his probability calculations yet again, found that he had been deceived by the factor 5. The improbability of our results was again reduced, though without reaching a degree which one could have described as probable. *The errors all tend to exaggerate the results in a way favourable to astrology,* and add most suspiciously to the impression of an artificial or fraudulent arrangement of the facts, which was so mortifying to those concerned that they would probably have preferred to keep silent about it.

I know, however, from long experience of these things that spontaneous synchronistic phenomena draw the observer, by hook or by crook, into what is happening and occasionally make him an accessory to the deed. That is the danger inherent in all parapsychological experiments. The dependence of ESP on an emotional factor in the experimenter and subject is a case in point. I therefore consider it a scientific duty to give as complete an account as possible of the result and to show how not only the statistical material, but the psychic processes of the interested parties, were affected by the synchronistic arrangement. Although, warned by previous experience, I was cautious enough to submit my original account (in the Swiss edition) to four competent persons, among them two mathematicians, I allowed myself to be lulled into a sense of security too soon.

The corrections made here do not in any way alter the fact that the maximal frequencies lie with the three classical lunar aspects.

In order to assure myself of the chance nature of the result, I undertook one more statistical experiment. I broke up the original and fortuitous chronological order and the equally fortuitous division into three batches by mixing the first 150 marriages with the last 150, taking the latter in reverse order; that is to say, I put the first marriage on top of the last, and then the second on top of the last but one, and so on. Then I divided the 300 marriages into three batches of a hundred. The result was as follows:

	1st Batch	2nd Batch	3rd Batch
Maximum	No Aspects 11%	☉ ☌ ♂ 11% ☾ ☌ ☾ 11%	☾ ☌ Asc. 12%

The result of the first batch is amusing in so far as only fifteen of the 300 marriages have none of the fifty selected aspects in common. The second batch yields two maxima, of which the second again represents a classical conjunction. The third batch yields a maximum for ☾ ☌ *Asc.*, which we already know as the third "classical" conjunction. The total result shows that another chance arrangement of the marriages can easily produce a result that deviates from the earlier total, but still does not quite prevent the classical conjunctions from turning up.

The result of our experiment tallies with our experience of mantic procedures. One has the impression that these methods, and others like them, create favourable conditions for the occurrence of meaningful coincidences. It is quite true that the verification of synchronistic phenomena is a difficult and sometimes impossible task. Rhine's achievement in demonstrating, with the help of unexceptionable material, the coincidence of a psychic state with a corresponding objective process must therefore be rated all the higher. Despite the fact that the statistical method is in general highly unsuited to do justice to unusual events, Rhine's experiments have nevertheless withstood the ruinous influence of statistics. Their results must therefore be taken into account in any assessment of synchronistic phenomena.

In view of the levelling influence which the statistical method has on the quantitative determination of synchronicity, we must ask how it was that Rhine succeeded in obtaining positive results. I maintain that he would never have got the results he did if he had carried out his experiments with a single subject,[13] or only a few. He needed a constant renewal of interest, an emotion with its characteristic *abaissement mental,* which tips the scales in favour of the unconscious. Only in this way can space and time be relativized to a certain extent, thereby reducing the chances of a causal process. What then happens

[13] By which I mean a subject chosen at random, and not one with specific gifts.

is a kind of *creatio ex nihilo,* an act of creation that is not causally explicable. The mantic procedures owe their effectiveness to this same connection with emotionality: by touching an unconscious aptitude they stimulate interest, curiosity, expectation, hope, and fear, and consequently evoke a corresponding preponderance of the unconscious. The effective (numinous) potencies in the unconscious are the archetypes. By far the greatest number of spontaneous synchronistic phenomena that I have had occasion to observe and analyse can easily be shown to have a direct connection with an archetype. This, in itself, is an irrepresentable, psychoid factor [14] of the collective unconscious. The latter cannot be localized, since it is either complete in principle in every individual or is found to be the same everywhere. You can never say with certainty whether what appears to be going on in the collective unconscious of a single individual is not also happening in other individuals or organisms or things or situations. When, for instance, the vision arose in Swedenborg's mind of a fire in Stockholm, there was a real fire raging there at the same time, without there being any demonstrable or even thinkable connection between the two.[15] I certainly would not like to undertake to prove the archetypal connection in this case. I would only point to the fact that in Swedenborg's biography there are certain things which throw a remarkable light on his psychic state. We must assume that there was a lowering of the

[14] Cf. "The Spirit of Psychology," in *Spirit and Nature* (Papers from the Eranos Yearbooks, 1; New York, 1954; London, 1955).

[15] This case is well authenticated. See report in Kant's "Dreams of a Spirit-Seer, Illustrated by Dreams of Metaphysics" (English trans., London, 1900).

threshold of consciousness which gave him access to "absolute knowledge." The fire in Stockholm was, in a sense, burning in him too. For the unconscious psyche space and time seem to be relative; that is to say, knowledge finds itself in a space-time continuum in which space is no longer space, nor time time. If, therefore, the unconscious should develop or maintain a potential in the direction of consciousness, it is then possible for parallel events to be perceived or "known."

Compared with Rhine's work the great disadvantage of my astrological statistics lies in the fact that the entire experiment was carried out on only one subject, myself. I did not experiment with a variety of subjects; rather, it was the varied material that challenged *my* interest. I was thus in the position of a subject who is at first enthusiastic, but afterwards cools off on becoming habituated to the ESP experiment. The results therefore deteriorated with the growing number of experiments, which in this case corresponded to the exposition of the material in batches, so that the accumulation of larger numbers only blurred the "favourable" initial result. Equally my final experiment showed that the discarding of the original order and the division of the horoscopes into arbitrary batches produce, as might be expected, a different picture, though its significance is not altogether clear.

Rhine's rules are to be recommended wherever (as in medicine) very large numbers are not involved. The interest and expectancy of the investigator might well be accompanied synchronistically by surprisingly favourable results to begin with, despite every precaution. These will be interpreted as "miracles" only by persons insufficiently acquainted with the statistical character of natural law.[16]

If—and it seems plausible—the meaningful coincidence or "cross-connection" of events cannot be explained causally, then the connecting principle must lie in the *equal significance* of parallel events; in other words, their *tertium comparationis* is *meaning*. We are so accustomed to regard meaning as a psychic process or content that it never enters our heads to suppose that it could also exist outside the psyche. But we do know at least enough about the psyche not to attribute to it any magical power, and still less can we attribute any magical power to the conscious mind. If, therefore, we entertain the hypothesis that one and the same (transcendental) meaning might manifest itself simultaneously in the human psyche and in the arrangement of an external and independent event, we at once come into conflict with the conventional scientific and epistemological views. We have to remind ourselves over and over again of the merely statistical validity of natural laws and of the effect of the statistical method in eliminating all unusual occurrences, if we want to lend an ear to such an hypothesis. The great difficulty is that we have absolutely no scientific means of proving the existence of an *objective* meaning which is not just a psychic product. We are, however, driven to some such assumption unless we want to regress to a *magical causality* and ascribe to the psyche a power that far exceeds its em-

[16] Cf. the interesting reflections of G. Spencer Brown: "De la recherche psychique considérée comme un test de la théorie des probabilités," *Revue métapsychique* (Paris), May–Aug., 1954, 87*ff*.

pirical range of action. In that case we should have to suppose, if we don't want to let causality go, either that Swedenborg's unconscious staged the Stockholm fire, or conversely that the objective event activated in some quite inconceivable manner the corresponding images in Swedenborg's brain. In either case we come up against the unanswerable question of transmission discussed above. It is of course entirely a matter of subjective opinion which hypothesis is felt to make more sense. Nor does tradition help us much in choosing between magical causality and transcendental meaning, because on the one hand the primitive mentality has always explained synchronicity as magical causality right down to our own day, and on the other hand philosophy assumed a secret correspondence or meaningful connection between natural events until well into the eighteenth century. I prefer the latter hypothesis because it does not, like the first, conflict with the empirical concept of causality, and can count as a principle *sui generis*. That obliges us, not indeed to correct the principles of natural explanation as hitherto understood, but at least to add to their number, an operation which only the most cogent reasons could justify. I believe, however, that the hints I have given in the foregoing constitute an argument that needs thorough consideration. Psychology, of all the sciences, cannot in the long run afford to overlook such experiences. These things are too important for an understanding of the unconscious, quite apart from their philosophical implications.

APPENDIX TO CHAPTER 2

[The following notes have been compiled by the Editors on the basis of Professor Fierz's mathematical argument, of which he kindly furnished a précis. These represent his latest thoughts on the topic. These data are presented here for the benefit of readers with a special interest in mathematics or statistics who want to know how the figures in the text were arrived at.

Since an orbit of 8° was taken as the basis of Professor Jung's calculations for the estimation of conjunctions and oppositions (cf. p. 64), it follows that, for a particular relation between two heavenly bodies to be called a conjunction (e.g., sun ☌ moon), one of them must lie within an arc of 16°. (Since the only concern was to test the character of the distribution, an arc of 15° was taken for convenience.)

Now, all positions on a circle of 360° are equally probable. So the probability α that the heavenly body will lie on an arc of 15° is

$$\alpha = \frac{15}{360} = \frac{1}{24} \qquad (1)$$

This probability α holds for every aspect.

Let n be the number of particular aspects that will occur in N married pairs if the probability that it will occur in one married pair be α.

Applying the binomial distribution, we get:

$$W_n = \frac{N!}{n!(N-n)!}\,\alpha^n(1-\alpha)^{N-n} \qquad (2)$$

In order to obtain a numerical evaluation of W_n, (2) can be simplified. This results in an error, which, however, is not important. The simplification can be arrived at by replacing (2) by the Poisson distribution:

$$P_n = \frac{1}{n!}\,x^n \cdot e^{-x}$$

This approximation is valid if α may be regarded as very small in comparison with 1, while x is finite.

Upon the basis of these considerations the following numerical results can be arrived at:

(a) The probability of ☾ ☌ ☉, ☾ ☌ ☾, and ☾ ☌ *Asc.* turning up simultaneously is:

$$\alpha^3 = \left(\frac{1}{24}\right)^3 \sim \frac{1}{10{,}000}$$

(b) The probability P for the maximal figures in the three batches is:

1. 18 aspects in 180 married pairs, $P = 1 : 1{,}000$
2. 24 aspects in 220 married pairs, $P = 1 : 10{,}000$
3. 8 aspects in 83 married pairs, $P = 1 : 50$.

—EDITORS.]

CHAPTER THREE

FORERUNNERS OF THE IDEA OF SYNCHRONICITY

The causality principle asserts that the connection between cause and effect is a necessary one. The synchronicity principle asserts that the terms of a meaningful coincidence are connected by *simultaneity* and *meaning*. So if we assume that the ESP experiments and numerous other observations are established facts, we must conclude that besides the connection between cause and effect there is another factor in nature which expresses itself in the arrangement of events and appears to us as meaning. Although meaning is an anthropomorphic interpretation it nevertheless forms the indispensable criterion of synchronicity. What that factor which appears to us as "meaning" may be in itself we have no possibility of knowing. As an hypothesis, however, it is not quite so impossible as may appear at first sight. We must remember that the rationalistic attitude of the West is not the only possible one and is not all-embracing, but is in many ways a prejudice and a bias that ought perhaps to be corrected. The very much older civilization of the Chinese has always thought differently from us in this respect, and we have to go back to Heraclitus if we want to find something similar in our civilization, at least where philosophy is concerned. Only in astrology, alchemy, and the mantic procedures do we find no differences of principle between our attitude and the Chinese. That is why alchemy developed along parallel lines in East and West and why in both ambits it strove towards the same goal with more or less identical ideas.[1]

In Chinese philosophy one of the oldest and most central ideas is that of Tao, which the Jesuits translated as "God." But that is correct only for the Western way of thinking. Other translations, such as "Providence" and the like, are mere makeshifts. Richard Wilhelm brilliantly interprets it as "meaning." [2] The concept of Tao pervades the whole philosophical thought of China. Causality occupies this paramount position with us, but it acquired its importance only in the course of the last two centuries, thanks to the levelling influence of the statistical method on the one hand and the unparalleled success of the natural sciences on the other, which brought the metaphysical view of the world into disrepute.

[1] Cf. my *Psychology and Alchemy* (London and New York, 1953), p. 343, and *Symbolik des Geistes* (Zurich, 1948), p. 115. Also the doctrine of *chen-yen* in Wei Po-yang ("An Ancient Chinese Treatise on Alchemy Entitled Ts'ang T'ung Ch'i," trans. by Lu-ch'iang Wu, *Isis*, Bruges, XVIII, 1932, 241, 251) and in Chuang-tzu.

[2] Wilhelm and Jung, *The Secret of the Golden Flower* (6th imp., London, 1945), p. 94, and Wilhelm, *Chinesische Lebensweisheit* (Darmstadt, 1922).

Lao-tzu gives the following description of Tao in his celebrated *Tao Teh Ching:* [3]

> There is something formless yet complete
> That existed before heaven and earth.
> How still! how empty!
> Dependent on nothing, unchanging,
> All pervading, unfailing.
> One may think of it as the mother of all things under heaven.
> I do not know its name,
> But I call it "Meaning."
> If I had to give it a name, I should call it "The Great."
> [Ch. XXV.]

Tao "covers the ten thousand things like a garment but does not claim to be master over them" (Ch. XXXIV). Lao-tzu describes it as "Nothing," [4] by which he means, says Wilhelm, only its "contrast with the world of reality." Lao-tzu describes its nature as follows:

> We put thirty spokes together and call it a wheel;
> But it is on the space where there is nothing that the utility of the wheel depends.
> We turn clay to make a vessel;
> But it is on the space where there is nothing that the utility of the vessel depends.
> We pierce doors and windows to make a house;
> And it is on these spaces where there is nothing that the utility of the house depends.
> Therefore just as we take advantage of what is, we should recognize the utility of what is not. [Ch. XI.]

"Nothing" is evidently "meaning" or "purpose," and it is only called Nothing because it does not manifest it-

[3] [Quotations from Arthur Waley's *The Way and Its Power* (London, 1934), with occasional slight changes to fit Wilhelm's reading.—Trans.]

[4] Tao is the contingent, which Andreas Speiser defines as "pure nothing" ("Über die Freiheit," *Basler Universitätsreden,* XXVIII, 1950).

self in the world of the senses, but is only its organizer.[5] Lao-tzu says:

> Because the eye gazes but can catch no glimpse of it,
> It is called elusive.
> Because the ear listens but cannot hear it,
> It is called the rarefied.
> Because the hand feels for it but cannot find it,
> It is called the infinitesimal. . . .
> These are called the shapeless shapes,
> Forms without form,
> Vague semblances.
> Go towards them, and you can see no front;
> Go after them, and you see no rear. [Ch. XIV.]

Wilhelm describes it as "a borderline conception lying at the extreme edge of the world of appearances." In it, the opposites "cancel out in non-discrimination," but are still potentially present. "These seeds," he continues, "point to something that corresponds firstly to *the visible,* i.e., something in the nature of an image; secondly to *the audible,* i.e., something in the nature of words; thirdly to *extension in space,* i.e., something with a form. But these three things are not clearly distinguished and definable, they are a non-spatial and non-temporal unity, having no above and below or front and back." As the *Tao Teh Ching* says:

> Incommensurable, impalpable,
> Yet latent in it are forms;
> Impalpable, incommensurable,
> Yet within it are entities.
> Shadowy it is and dim. [Ch. XXI.]

[5] Wilhelm, *Chinesische Lebensweisheit,* p. 15: "The relation between meaning (Tao) and reality cannot be conceived, either, under the category of cause and effect."

Reality, thinks Wilhelm, is conceptually knowable because according to the Chinese view there is in all things a latent "rationality." [6] This is the basic idea underlying meaningful coincidence: it is possible because both sides have the same meaning. Where meaning prevails, order results:

> Tao is eternal, but has no name;
> The Uncarved Block, though seemingly of small account,
> Is greater than anything under heaven.
> If the kings and barons would but possess themselves of it,
> The ten thousand creatures would flock to do them homage;
> Heaven and earth would conspire
> To send Sweet Dew;
> Without law or compulsion men would dwell in harmony.
> [Ch. XXXII.]
>
> Tao never does;
> Yet through it all things are done. [Ch. XXXVII.]
>
> Heaven's net is wide;
> Coarse are the meshes, yet nothing slips through.
> [Ch. LXXIII.]

Chuang-tzu (a contemporary of Plato's) says of the psychological premises on which Tao is based: "The state in which ego and non-ego are no longer opposed is called the pivot of Tao." [7] It sounds almost like a criticism of our scientific view of the world when he remarks that "Tao is obscured when you fix your eye on little segments of existence only" [8] or "Limitations are not originally

[6] Ibid., p. 19.

[7] *Das wahre Buch vom südlichen Blütenland,* trans. by R. Wilhelm (Jena, 1912), II, 3.

[8] Ibid., II, 3.

grounded in the meaning of life. Originally words had no fixed meanings. Differences only arose through looking at things subjectively." [9] The sages of old, says Chuang-tzu, "took as their starting-point a state when the existence of things had not yet begun. That is indeed the extreme limit beyond which you cannot go. The next assumption was that though things existed they had not yet begun to be separated. The next, that though things were separated in a sense, affirmation and negation had not yet begun. When affirmation and negation came into being, Tao faded. After Tao faded, then came one-sided attachments." [10] "Outward hearing should not penetrate further than the ear; the intellect should not seek to lead a separate existence, thus the soul can become empty and absorb the whole world. It is Tao that fills this emptiness." If you have insight, says Chuang-tzu, "you use your inner eye, your inner ear, to pierce to the heart of things, and have no need of intellectual knowledge." [11] This is obviously an allusion to the absolute knowledge of the unconscious, and to the presence in the microcosm of macrocosmic events.

This Taoistic view is typical of Chinese thinking. It is, whenever possible, *a thinking in terms of the whole,* a point also brought out by Marcel Granet,[12] the eminent authority on Chinese psychology. This peculiarity can be

[9] II, 7.

[10] II, 5.

[11] IV, 1.

[12] *La Pensée chinoise* (Paris, 1934); also Lily Abegg, *The Mind of East Asia* (London and New York, 1952). The latter gives an excellent account of the synchronistic mentality of the Chinese.

seen in ordinary conversation with the Chinese: what seems to us a perfectly straightforward, precise question about some detail evokes from the Chinese thinker an unexpectedly elaborate answer, as though one had asked him for a blade of grass and got a whole meadow in return. With us details are important for their own sakes; for the Oriental mind they always complete a total picture. In this totality, as in primitive or in our own medieval, pre-scientific psychology (still very much alive!), are included things which seem to be connected with one another only "by chance," by a coincidence whose meaningfulness appears altogether arbitrary. This is where the theory of *correspondentia*[13] comes in, which was propounded by the natural philosophers of the Middle Ages, and particularly the classical idea of the *sympathy of all things.*[14] Hippocrates says:

> There is one common flow, one common breathing, all things are in sympathy. The whole organism and each one of its parts are working in conjunction for the same purpose . . . the great principle extends to the extremest part, and from the extremest part it returns to the great principle, to the one nature, being and not-being.[15]

[13] Professor W. Pauli kindly calls my attention to the fact that Niels Bohr used "correspondence" as a mediating term between the representation of the discontinuum (particle) and the continuum (wave). Originally (1913–18) he called it the "principle of correspondence," but later (1927) it was formulated as the "argument of correspondence."

[14] "συμπάθεια τῶν ὅλων."

[15] "De alimento," a tract ascribed to Hippocrates. (Trans. by John Precope in *Hippocrates on Diet and Hygiene,* London, 1952, p. 174, modified.) "Σύρροια μία, συμπνοία μία, πάντα συμπαθέα κατὰ

The universal principle is found even in the smallest particle, which therefore corresponds to the whole.

In this connection there is an interesting idea in Philo (25 B.C.–A.D. 42):

> God, being minded to unite in intimate and loving fellowship the beginning and end of created things, made heaven the beginning and man the end, the one the most perfect of imperishable objects of sense, the other the noblest of things earthborn and perishable, being, in very truth, a miniature heaven. He bears about within himself, like holy images, endowments of nature that correspond to the constellations. . . . For since the corruptible and the incorruptible are by nature contrary the one to the other, God assigned the fairest of each sort to the beginning and the end, heaven (as I have said) to the beginning, and man to the end.[16]

Here the great principle [16a] or beginning, heaven, is infused into man the microcosm, who reflects the star-like natures and thus, as the smallest part and end of the work of Creation, contains the whole.

According to Theophrastus (371–288 B.C.) the supersensuous and the sensuous are joined by a bond of community. This bond cannot be mathematics, so must presumably be God.[17] Similarly in Plotinus the individual souls born of the one World Soul are related to one an-

μὲν οὐλομελίην πάντα κατὰ μέρος δὲ τὰ ἐν ἑκάστῳ μέρει μερέα πρὸς τὸ ἔργον . . . ἀρχὴ μεγάλη ἐς ἔσχατον μέρος ἀφικνέεται, ἐξ ἐσχάτου μέρεος εἰς ἀρχὴν μεγάλην ἀφικνέεται, μία φύσις εἶναι καὶ μὴ εἶναι."

[16] *De opificio mundi,* 82 (trans. by F. H. Colson and G. H. Whitaker in the Loeb Classical Library edn. of Philo, London and Cambridge, Mass., I, 1929, p. 67).

[16a] "ἀρχὴ μεγάλη"

[17] Eduard Zeller, *Die Philosophie der Griechen* (Tübingen, 1856), II, part ii, p. 654.

other by sympathy or antipathy, regardless of distance.[18] Similar views are to be found in Pico della Mirandola:

> Firstly there is the unity in things whereby each thing is at one with itself, consists of itself, and coheres with itself. Secondly there is the unity whereby one creature is united with the others and all parts of the world constitute one world. The third and most important (unity) is that whereby the whole universe is one with its Creator, as an army with its commander.[19]

By this threefold unity Pico means a simple unity which, like the Trinity, has three aspects; "a unity distinguished by a threefold character, yet in such a way as not to depart from the simplicity of unity."[20] For him the world is *one* being, a visible God, in which everything is naturally arranged from the very beginning like the parts of a living organism. The world appears as the *corpus mysticum* of God, just as the Church is the *corpus mysticum* of Christ, or as a well-disciplined army can be called a sword in the hand of the commander. The view that all things are arranged according to God's will is one that leaves little room for causality. Just as in a living body the different parts work in harmony and are meaningfully adjusted to one another, so events in the world stand in a meaningful relationship which cannot be derived from

[18] *Enneads,* IV, 3, 8 and 4, 32 (in A. C. H. Drews, *Plotin und der Untergang der antiken Weltanschauung,* Jena, 1907, p. 179).

[19] *Heptaplus,* VI, prooem., in *Opera omnia* (Basel, 1557), pp. 40 *f.* ("Est enim primum ea in rebus unitas, qua unumquodque sibi est unum sibique constat atque cohaeret. Est ea secundo, per quam altera alteri creatura unitur, et per quam demum omnes mundi partes unus sunt mundus. Tertia atque omnium principalissima est, qua totum universum cum suo opifice quasi exercitus cum suo duce est unum.")

[20] "unitas ita ternario distincta, ut ab unitatis simplicitate non discedat."

any immanent causality. The reason for this is that in either case the behaviour of the parts depends on a central control which is superordinate to them.

In his treatise *De hominis dignitate* Pico says: "The Father implanted in man at birth seeds of all kinds and the germs of original life." [21] Just as God is the "copula" of the world, so, within the created world, is man. "Let us make man in our image, who is not a fourth world or anything like a new nature, but is rather the fusion and synthesis of three worlds (the supercelestial, the celestial, and the sublunary)." [22] In body and spirit man is "the little God of the world," the microcosm.[23] Like God, therefore, man is a centre of events, and all things revolve about him.[24] This thought, so utterly strange to the modern mind, dominated man's picture of the world until a few generations ago, when natural science proved man's subordination to nature and his extreme dependence on causes. The idea of a correlation between events and meaning (now assigned exclusively to man) was banished to such a remote and benighted region that the intellect

[21] *Opera omnia,* p. 315. ("Nascenti homini omnifaria semina et origenae vitae germina indidit pater.")

[22] *Heptaplus,* V, vi, in ibid., p. 38. ("Faciamus hominem ad imaginem nostram, qui non tam quartus est mundus, quasi nova aliqua natura, quam trium (mundus supercoelestis, coelestis, sublunaris) complexus et colligatio.")

[23] "God . . . placed man in the centre [of the world] after his image and the similitude of forms" ("Deus . . . hominem in medio [mundi] statuit ad imaginem suam et similitudinem formarum").

[24] Pico's doctrine is a typical example of the medieval correspondence theory. A good account of cosmological and astrological correspondence is to be found in Alfons Rosenberg, *Zeichen am Himmel: Das Weltbild der Astrologie* (Zurich, 1949).

lost track of it altogether. Schopenhauer remembered it somewhat belatedly after it had formed one of the chief items in Leibnitz's scientific explanations.

By virtue of his microcosmic nature man is a son of the firmament or macrocosm. "I am a star travelling together with you," the initiate confesses in the Mithraic liturgy.[25] In alchemy the microcosmos has the same significance as the *rotundum,* a favourite symbol since the time of Zosimos of Panopolis, which was also known as the Monad.

The idea that the inner and outer man together form the whole, the οὐλομελίη of Hippocrates, a microcosm or smallest part wherein the "great principle" is undividedly present, also characterizes the thought of Agrippa von Nettesheim. He says:

> It is the unanimous consent of all Platonists, that as in the archetypal World, all things are in all; so also in this corporeal world, all things are in all, albeit in different ways, according to the receptive nature of each. Thus the Elements are not only in these inferiour bodies, but also in the Heavens, in Stars, in Divels, in Angels, and lastly in God, the maker, and archetype of all things.[26]

[25] Albrecht Dieterich, *Eine Mithrasliturgie* (Leipzig, 1903), p. 9.

[26] Henricus Cornelius Agrippa von Nettesheim, *De occulta philosophia Libri tres* (Cologne, 1533), I, viii, p. 12. Trans. by "J. F." as *Three Books of Occult Philosophy* (London, 1651), p. 20; republished under the editorship of W. F. Whitehead (Chicago, 1898; Book I only), p. 55. [Quotations from the J. F. translation have been slightly modified.—Trans.] ("Est Platonicorum omnium unanimis sententia quemadmodum in archetypo mundo omnia sunt in omnibus, ita etiam in hoc corporeo mundo, omnia in omnibus esse, modis tamen diversis, pro natura videlicet suscipientium: sic et elementa non solum sunt in istis inferioribus, sed

The ancients had said: "All things are full of gods." [27] These gods were "divine powers which are diffused in things." [28] Zoroaster had called them "divine allurements," [29] and Synesius "symbolic inticements." [30] This latter interpretation comes very close indeed to the idea of archetypal projections in modern psychology, although from the time of Synesius until quite recently there was no epistemological criticism, let alone the newest form of it, namely psychological criticism. Agrippa shares with the Platonists the view that there is an "immanent power in the things of the lower world which makes them agree to a large extent with the things of the upper world," and that as a result the animals are connected with the "divine bodies" (i.e., the stars) and exert an influence on them.[31] Here he quotes Virgil: "I for my part do not believe that they [the rooks] are endowed with divine spirit or with a foreknowledge of things greater than the oracle." [32]

in coelis, in stellis, in daemonibus, in angelis, in ipso denique omnium opifice et archetypo.")

[27] "Omnia plena diis esse."

[28] "virtutes divinae in rebus diffusae"

[29] "divinae illices"

[30] "symbolicae illecebrae." [In J. F. original edn., p. 32; Whitehead edn., p. 69.—TRANS.] Agrippa is basing himself here on the Marsilio Ficino translation (*Auctores Platonici*, Venice, 1497, II, v°). In Synesius (*Opuscula*, ed. by Nicolaus Terzaghi [Scriptores Graeci et Latini], Rome, 1949, p. 148), the text of Περὶ ἐνυπνίων III B has τὸ θελγόμενον, from θέλγειν, "to excite, charm, enchant."

[31] *De occulta philosophia*, lv, p. 69. (J. F. edn., p. 117; Whitehead edn., p. 169.) Similarly in Paracelsus.

[32] "Haud equidem credo, quia sit divinius illis
Ingenium aut rerum fato prudentia maior."
—*Georgics*, I, 415 *f.*

Agrippa is thus suggesting that there is an inborn "knowledge" or "imagination" in living organisms, an idea which recurs in our own day in Hans Driesch.[33] Whether we like it or not, we find ourselves in this embarrassing position as soon as we begin seriously to reflect on the teleological processes in biology or to investigate the compensatory function of the unconscious, not to speak of trying to explain the phenomenon of synchronicity. Final causes, twist them how we will, postulate a *foreknowledge of some kind.* It is certainly not a knowledge that could be connected with the ego, and hence not a conscious knowledge as we know it, but rather a self-subsistent "unconscious" knowledge which I would prefer to call "absolute knowledge." It is not cognition but, as Leibnitz so excellently calls it, a "perceiving" which consists—or to be more cautious, seems to consist—of images, of subjectless "simulacra." These postulated images are presumably the same as my archetypes, which can be shown to be formal factors in spontaneous fantasy products. Expressed in modern language, the microcosm which contains "the images of all creation" would be the collective unconscious.[34] By the *spiritus mundi,* the *ligamentum animae et corporis,* the *quinta essentia,*[35] which he shares with the

[33] *Die "Seele" als elementarer Naturfaktor* (Leipzig, 1903), pp. 80, 82.

[34] Cf. "The Spirit of Psychology," in *Spirit and Nature* (Papers from the Eranos Yearbooks, 1; New York, 1954; London, 1955).

[35] Agrippa says of this (op. cit., I, xiv, p. 29; J. F. edn., p. 33; Whitehead edn., p. 70): "That which we call the quintessence: because it is not from the four Elements, but a certain fifth thing, having its being above, and besides them." ("Quoddam quintum super illa [elementa] aut praeter illa subsistens.")

alchemists, Agrippa probably means what we would call the unconscious. The spirit that "penetrates all things," or shapes all things, is the World Soul: "The soul of the world therefore is a certain only thing, filling all things, bestowing all things, binding, and knitting together all things, that it might make one frame of the world. . . ." [36] Those things in which this spirit is particularly powerful therefore have a tendency to "beget their like," [37] in other words, to produce correspondences or meaningful coincidences.[38] Agrippa gives a long list of these correspondences, based on the numbers 1 to 12.[39] A similar but more alchemical table of correspondences can be found in a treatise of Aegidius de Vadis.[40] Of these I would only mention the *scala unitatis,* because it is especially interesting from the point of view of the history of symbols: "Iod [the first letter of the tetragrammaton, the divine name]—anima mundi—sol—lapis

[36] II, lvii, p. 203 (J. F. edn., p. 331): "Est itaque anima mundi, vita quaedam unica omnia replens, omnia perfundens, omnia colligens et connectens, ut unam reddat totius mundi machinam. . . ."

[37] Ibid.: ". . . potentius perfectiusque agunt, tum etiam promptius generant sibi simile."

[38] The zoologist A. C. Hardy reaches similar conclusions: "Perhaps our ideas on evolution may be altered if something akin to telepathy—unconscious no doubt—were found to be a factor in moulding the patterns of behaviour among members of a species. If there was such a non-conscious group-behaviour plan, distributed between, and linking, the individuals of the race, we might find ourselves coming back to something like those ideas of subconscious racial memory of Samuel Butler, but on a group rather than an individual basis." *Discovery* (London), X (1949: Oct.), 328.

[39] Op. cit., II, iv–xiv.

[40] "Dialogus inter naturam et filium philosophiae," *Theatrum chemicum* (Ursel, 1602), II, p. 123.

philosophorum—cor—Lucifer." [41] I must content myself with saying that this is an attempt to set up a hierarchy of archetypes, and that tendencies in this direction can be shown to exist in the unconscious.[42]

Agrippa was an older contemporary of Theophrastus Paracelsus and is known to have had a considerable influence on him.[43] So it is not surprising if the thinking of Paracelsus proves to be steeped in the idea of correspondence. He says:

If a man will be a philosopher without going astray, he must lay the foundations of his philosophy by making heaven and earth a microcosm, and not be wrong by a hair's breadth. Therefore he who will lay the foundations of medicine must also guard against the slightest error, and must make from the microcosm the revolution of heaven and earth, so that the philosopher does not find anything in heaven and earth which he does not also find in man, and the physician does not find anything in man which heaven and earth do not have. And these two differ only in outward form, and yet the form on both sides is understood as pertaining to one thing.[44]

The *Paragranum* [45] has some pointed psychological remarks to make about physicians:

[41] Cited in Agrippa, op. cit., II, iv, p. 104 (J. F. edn., p. 176).

[42] Cf. Aniela Jaffé, "Bilder und Symbole aus E. T. A. Hoffmann's Märchen 'Der goldene Topf,'" in her and my *Gestaltungen des Unbewussten* (Zurich, 1950), and Marie-Louise von Franz, "Die Passio Perpetuae," in her and my *AION, Untersuchungen zur Symbolgeschichte* (Zurich, 1951).

[43] Cf. my *Paracelsica* (Zurich, 1942), pp. 47 *ff*.

[44] *Das Buch Paragranum,* ed. by Franz Strunz (Leipzig, 1903), pp. 35 *f*. Much the same in *Labyrinthus medicorum,* in the *Sämtliche Werke,* ed. by Karl Sudhoff (Munich and Berlin, 1922–33), XI, pp. 204 *ff*.

[45] Strunz edn., p. 34.

For this reason, [we assume] not four, but one arcanum, which is, however, four-square, like a tower facing the four winds. And as little as a tower may lack a corner, so little may the physician lack one of the parts. . . . At the same [time he] knows how the world is symbolized [by] an egg in its shell, and how a chick with all its substance lies hidden within it. Thus everything in the world and in man must lie hidden in the physician. And just as the hens, by their brooding, transform the world prefigured in the shell into a chick, so Alchemy brings to maturity the philosophical arcana lying in the physician. . . . Herein lies the error of those who do not understand the physician aright.[46]

What this means for alchemy I have shown in some detail in my *Psychology and Alchemy.*

Johannes Kepler thought in much the same way. He says in his *Tertius interveniens* (1610): [47]

This [viz., a geometrical principle underlying the physical world] is also, according to the doctrine of Aristotle, the strongest tie that links the lower world to the heavens and unifies it therewith so that all its forms are governed from on high; for in this lower world, that is to say the globe of the earth, there is inherent a spiritual nature, capable of *Geometria,* which *ex instinctu creatoris, sine ratiocinatione* comes to life and stimulates itself into a use of its forces through the geometrical and harmonious combination of the heavenly rays of light. Whether all plants and animals as well as the globe of the earth have this faculty in themselves I cannot say. But it is not an unbelievable thing. . . . For, in all these things [e.g., in the fact that flowers have a definite colour, form, and number of petals] there is at work the *instinctus divinus, rationis*

[46] Similar ideas in Jakob Böhme, *The Signature of All Things,* trans. by John Ellistone and ed. by Clifford Bax (Everyman's Library, London, 1912), p. 10: "Man has indeed the forms of all the three worlds in him, for he is a complete image of God, or of the Being of all beings. . . ." (*De signatura rerum,* Amsterdam, 1635, I, 7.)

[47] *Joannis Kepleri Astronomi Opera omnia,* ed. by C. Frisch (Frankfort on the Main and Erlangen, 1858 *ff.*), I, pp. 605 *ff.*

particeps, and not at all man's own intelligence. That man, too, through his soul and its lower faculties, has a like affinity to the heavens as has the soil of the earth can be tested and proven in many ways.[48]

Concerning the astrological "Character," i.e., astrological synchronicity, Kepler says:

> This *Character* is received, not into the body, which is much too inappropriate for this, but into the soul's own nature, which behaves like a point (for which reason it can also be transformed into the point of the *confluxus radiorum*). This [nature of the soul] not only partakes of their reason (on account of which we human beings are called reasonable above other living creatures) but also has another, innate reason [enabling it] to apprehend instantaneously, without long learning, the *Geometriam* in the *radiis* as well as in the *vocibus,* that is to say, in *Musica.*[49]
>
> Thirdly, another marvellous thing is that the nature which receives this *Characterem* also induces a certain correspondence *in constellationibus coelestibus* in its relatives. When a mother is great with child and the natural time of delivery is near, nature selects for the birth a day and hour which correspond, on account of the heavens [scil., from an astrological point of view], to the nativity of the mother's brother or father, and this *non qualitative, sed astronomice et quantitative.*[50]
>
> Fourthly, so well does each nature know not only its *characterem coelestem* but also the celestial *configurationes* and courses of every day that, whenever a planet moves *de praesenti* into its *characteris ascendentem* or *loca praecipua,* especially into the *Natalitia,*[50a] it responds to this and is affected and stimulated thereby in various ways.[51]

---

[48] Ibid., No. 64.

[49] No. 65. [50] No. 67.

[50a] ["in die Natalitia" = "into those [positions presiding] at birth," if "in die" is construed as German. The *Gesammelte Werke,* ed. by M. Caspar and F. Hammer (Munich, 1941), IV, p. 211, has "in die Natalitio" = "in the day of birth," the words "in die" being construed as Latin.—TRANS.]

[51] No. 68.

Kepler supposes that the secret of the marvellous correspondence is to be found in the *earth,* because the earth is animated by an *anima telluris,* for whose existence he adduces a number of proofs. Among these are: the constant temperature below the surface of the earth; the peculiar power of the earth-soul to produce metals, minerals, and fossils, namely the *facultas formatrix,* which is similar to that of the womb and can bring forth in the bowels of the earth shapes that are otherwise found only outside—ships, fishes, kings, popes, monks, soldiers, etc.; [52] further the practice of geometry, for it produces the five geometrical bodies and the six-cornered figures in crystals. The *anima telluris* has all this from an original impulse, independent of the reflection and ratiocination of man.[53]

The seat of astrological synchronicity is not in the planets but in the earth; [54] not in matter, but in the *anima telluris.* Therefore every kind of natural or living power in bodies has a certain "divine similitude." [55]

Such was the intellectual background when Gottfried Wilhelm von Leibnitz (1646–1716) appeared with his idea

[52] See the dreams mentioned below.

[53] Kepler, *Opera,* ed. by Frisch, V, p. 254; cf. also II, pp. 270 *f.* and VI, pp. 178 *f.* ". . . formatrix facultas est in visceribus terrae, quae feminae praegnantis more occursantes foris res humanas veluti eas videret, in fissibilibus lapidibus exprimit, ut militum, monachorum, pontificum, regum et quidquid in ore hominum est. . . ."

[54] ". . . quod scl. principatus causae in terra sedeat, non in planetis ipsis." Ibid., II, p. 642.

[55] ". . . ut omne genus naturalium vel animalium facultatum in corporibus Dei quandam gerat similitudinem." Ibid. I am indebted to Dr. Liliane Frey-Rohn and Dr. Marie-Louise von Franz for this reference to Kepler.

of *pre-established harmony,* that is, an absolute synchronism of psychic and physical events. This theory finally petered out in the concept of "psychophysical parallelism." Leibnitz's pre-established harmony and the above-mentioned idea of Schopenhauer's, that the unity of the primal cause produces a simultaneity and interrelationship of events not in themselves causally connected, are at bottom only a repetition of the old peripatetic view, with a modern deterministic colouring in the case of Schopenhauer and a partial replacement of causality by an antecedent order in the case of Leibnitz. For him God is the creator of order. He compares soul and body to two synchronized clocks [56]

[56] G. W. Leibnitz, "Second Explanation of the System of the Communication between Substances" (*The Philosophical Works of Leibniz,* a selection trans. by G. M. Duncan, New Haven, 1890, pp. 90–91): "From the beginning God has made each of these two substances of such a nature that merely by following its own peculiar laws, received with its being, it nevertheless accords with the other, just as if there were a mutual influence or as if God always put his hand thereto in addition to his general co-operation."

As Professor Pauli has kindly pointed out, it is possible that Leibnitz took his idea of the synchronized clocks from the Flemish philosopher Arnold Geulincx (1625–99). In his *Metaphysica vera, Part III,* there is a note to "Octava scientia" (*Opera philosophica,* The Hague, 1892, Vol. II, p. 195), which says (p. 296): ". . . horologium voluntatis nostrae quadret cum horologio motus in corpore" (the clock of our will is synchronized with the clock of our physical movement). Another note (p. 297) explains: "Voluntas nostra nullum habet influxum, causalitatem, determinationem aut efficaciam quamcunque in motum . . . cum cogitationes nostras bene excutimus, nullam apud nos invenimus ideam seu notionem determinationis. . . . Restat igitur Deus solus primus motor et solus motor, quia et ita motum ordinat atque disponit et ita simul voluntati nostrae licet libere moderatur, ut eodem temporis momento conspiret et voluntas nostra ad projiciendum v.g. pedes inter ambulandum,

and uses the same simile to express the relations of the monads or entelechies with one another. Although the monads cannot influence one another directly because, as he says, they "have no windows" [57] (relative abolition of causality!), they are so constituted that they are always in accord without having knowledge of one another. He conceives each monad to be a "little world" or "active indivisible mirror." [58] Not only is man a microcosm enclosing the whole in himself, but every entelechy or monad is in effect such a microcosm. Each "simple substance"

et simul ipsa illa pedum projectio seu ambulatio." (Our will has no influence, no causative or determinative power, and no effect of any kind on our movement. . . . If we examine our thoughts carefully, we find in ourselves no idea or concept of determination. . . . There remains, therefore, only God as the prime mover and only mover, because he arranges and orders movement and freely co-ordinates it with our will, so that our will wishes simultaneously to throw the feet forward into walking, and simultaneously the forward movement and the walking takes place.) A note to "Nona scientia" adds (p. 298): "Mens nostra . . . penitus independens est ab illo (scl. corpore) . . . omnia quae de corpore scimus jam praevie quasi ante nostram cognitionem esse in corpore. Ut illa quodam modo nos in corpore legamus, non vero inscribamus, quod Deo proprium est." (Our mind . . . is totally independent of the body . . . everything we know about the body is already in the body, before our thought. So that we can, as it were, read ourselves in our body, but not imprint ourselves on it. Only God can do that.) This idea anticipates Leibnitz' clock comparison.

[57] *Monadology,* § 7 (*Leibniz: Philosophical Writings,* selected and trans. by Mary Morris, Everyman's Library, London, 1934, p. 3): "Monads have no windows, by which anything could come in or go out. . . . Thus neither substance nor accident can enter a monad from without."

[58] Rejoinder to the remarks in Bayle's Dictionary, from the *Kleinere philosophische Schriften* (ed. by R. Habs, Leipzig, 1883), XI, p. 105.

has connections "which express all the others." It is "a perpetual living mirror of the universe." [59] He calls the monads of living organisms "souls": "the soul follows its own laws, and the body its own likewise, and they accord by virtue of the harmony pre-established among all substances, since they are all representations of one and the same universe." [60] This clearly expresses the idea that man is a microcosm. "Souls in general," says Leibnitz, "are the living mirrors or images of the universe of created things." He distinguishes between minds on the one hand, which are "images of the Divinity . . . capable of knowing the system of the universe, and of imitating something of it by architectonic patterns, each mind being as it were a little divinity in its own department," [61] and bodies on the other hand, which "act according to the laws of efficient causes by motions," while the souls act "according to the laws of final causes by appetitions, ends, and means." [62] In the monad or soul alterations take place whose cause is the "appetition." [63] "The passing state, which involves and represents a plurality within the unity or simple substance, is nothing other than what is called perception," says Leibnitz.[64] Perception is the "inner state

[59] *Monadology,* § 56 (Morris edn., p. 12): "Now this connection or adaptation of all created things with each, and of each with all the rest, means that each simple substance has relations which express all the others, and that consequently it is a perpetual living mirror of the universe."

[60] Ibid., § 78 (p. 17).

[61] § 83 (p. 18); cf. *Theodicy,* § 147 (trans. by E. M. Huggard, ed. by Austin Farrer, New Haven, 1952, pp. 215 *f.*).

[62] *Monadology,* § 79 (Morris edn., p. 17).

[63] Ibid., § 15 (p. 5).

[64] § 14 (pp. 4 *f.*).

of the monad representing external things," and it must be distinguished from conscious apperception. "For perception is unconscious." [65] Herein lay the great mistake of the Cartesians, "that they took no account of perceptions which are not apperceived." [66] The perceptive faculty of the monad corresponds to the *knowledge,* and its appetitive faculty to the *will,* that is in God.[67]

It is clear from these quotations that besides the causal connection Leibnitz postulates a complete pre-established parallelism of events both inside and outside the monad. The synchronicity principle thus becomes the absolute rule in all cases where an inner event occurs simultaneously with an outside one. As against this, however, it must be borne in mind that the synchronistic phenomena which can be verified empirically, far from constituting a rule, are so exceptional that most people doubt their existence. They certainly occur much more frequently in reality than one thinks or can prove, but we still do not know whether they occur so frequently and so regularly in any field of experience that we could speak of them as conforming to law.[68] We only know that there must be an underlying

[65] "Principles of Nature and of Grace, Founded on Reason," § 4 (Morris edn., p. 22).

[66] *Monadology,* § 14 (p. 5). Cf. also Dr. Marie-Louise von Franz's paper on the dream of Descartes in *Zeitlose Dokumente der Seele* (Studien aus dem C. G. Jung Institut, III; Zurich, 1952).

[67] *Monadology,* § 48 (p. 11); *Theodicy,* § 149.

[68] I must again stress the possibility that the relation between body and soul may yet be understood as a synchronistic one. Should this conjecture ever be proved, my present view that synchronicity is a relatively rare phenomenon would have to be corrected. Cf. C. A. Meier's observations in *Zeitgemässe Probleme der Traumforschung* (Eidgenössische Technische Hochschule:

principle which might possibly explain all such (related) phenomena.

The primitive as well as the classical and medieval views of nature postulate the existence of some such principle alongside causality. Even in Leibnitz, causality is neither the only view nor the predominant one. Then, in the course of the eighteenth century, it became the exclusive principle of natural science. With the rise of the physical sciences in the nineteenth century the correspondence theory vanished completely from the surface, and the magical world of earlier ages seemed to have disappeared once and for all until, towards the end of the century, the founders of the Society for Psychical Research indirectly opened up the whole question again through their investigation of telepathic phenomena.

The medieval attitude of mind I have described above underlies all the magical and mantic procedures which have played an important part in man's life since the remotest times. The medieval mind would regard Rhine's laboratory-arranged experiments as magical performances, whose effect for this reason would not seem so very astonishing. It was interpreted as a "transmission of energy," which is still commonly the case today, although, as I have said, it is not possible to form any empirically verifiable conception of the transmitting medium.

I need hardly point out that for the primitive mind synchronicity is a self-evident fact; consequently at this stage there is no such thing as chance. No accident, no illness, no death is ever fortuitous or attributable to

Kultur- und Staatswissenschaftliche Schriften, 75; Zurich, 1950), p. 22.

"natural" causes. Everything is somehow due to magical influence. The crocodile that catches a man while he is bathing has been sent by a magician; illness is caused by some spirit or other; the snake that was seen by the grave of somebody's mother is obviously her soul; etc. On the primitive level, of course, synchronicity does not appear as an idea by itself, but as "magical" causality. This is an early form of our classical idea of causality, while the development of Chinese philosophy produced from the connotation of the magical the "concept" of Tao, of meaningful coincidence, but no causality-based science.

Synchronicity postulates a meaning which is *a priori* in relation to human consciousness and apparently exists outside man.[69] Such an assumption is found above all in the philosophy of Plato, which takes for granted the existence of transcendental images or models of empirical things, the εἴδη (forms, species), whose reflections (εἴδωλα) we see in the phenomenal world. This assumption not only presented no difficulty to earlier centuries but was on the contrary perfectly self-evident. The idea of an *a priori* meaning may also be found in the older mathematics, as in the mathematician Jacobi's paraphrase of Schiller's poem "Archimedes and His Pupil." He praises the calculation of the orbit of Uranus and closes with the lines:

> What you behold in the cosmos is only the light of God's glory;
> In the Olympian host Number eternally reigns.

[69] In view of the possibility that synchronicity is not only a psychophysical phenomenon but might also occur without the participation of the human psyche, I should like to point out that in this case we should have to speak not of *meaning* but of equivalence or conformity.

The great mathematician Gauss is the putative author of the saying: "God does arithmetic." [70]

The idea of synchronicity and of a self-subsistent meaning, which forms the basis of classical Chinese thinking and of the naïve views of the Middle Ages, seems to us an archaic assumption that ought at all costs to be avoided. Though the West has done everything possible to discard this antiquated hypothesis, it has not quite succeeded. Certain mantic procedures seem to have died out, but astrology, which in our own day has attained an eminence never known before, remains very much alive. Nor has the determinism of a scientific epoch been able to extinguish altogether the persuasive power of the synchronicity principle. For in the last resort it is not so much a question of superstition as of a truth which remained hidden for so long only because it had less to do with the physical side of events than with their psychic aspects. It was modern psychology and parapsychology which proved that causality does not explain a certain class of events and that in this case we have to consider a formal factor, namely synchronicity, as a principle of explanation.

For those who are interested in psychology I should like to mention here that the peculiar idea of a self-sub-

[70] "ὁ θεὸς ἀριθμητίζει." But in a letter of 1830 Gauss says: "We must in all humility admit that if number is *merely* a product of our mind, space has a reality outside our mind." (Leopold Kronecker, *Über den Zahlenbegriff*, in his *Werke*, III, 1899, p. 252.) Hermann Weyl likewise takes number as a product of reason. ("Wissenschaft als symbolische Konstruktion des Menschen," *Eranos-Jahrbuch 1948*, p. 375). Markus Fierz, on the other hand, inclines more to the Platonic idea. ("Zur physikalischen Erkenntnis," *Eranos-Jahrbuch 1948*, p. 434.)

sistent meaning is suggested in dreams. Once when this idea was being discussed in my circle somebody remarked: "The geometrical square does not occur in nature except in crystals." A lady who had been present had the following dream that night: *In the garden there was a large sandpit in which layers of rubbish had been deposited. In one of these layers she discovered thin, slaty plates of green serpentine. One of them had black squares on it, arranged concentrically. The black was not painted on, but was ingrained in the stone, like the markings in an agate. Similar marks were found on two or three other plates, which Mr. A (a slight acquaintance) then took away from her.*[71] Another dream motif of the same kind is the following: *The dreamer was in a wild mountain region where he found contiguous layers of triassic rock. He loosened the slabs and discovered to his boundless astonishment that they had human heads on them in low relief.* This dream was repeated several times at long intervals.[72] Another time the dreamer *was travelling through the Siberian tundra and found an animal he had long been looking for. It was a more than lifesize cock, made of what looked like thin, colourless glass. But it was alive and had just sprung by chance from a microscopic unicellular organism which had the power to turn into all sorts of animals (not otherwise found in the tundra) or even into objects*

[71] According to the rules of dream interpretation this Mr. A would represent the animus, who, as a personification of the unconscious, takes back the designs because the conscious mind has no use for them and regards them only as *lusus naturae*.

[72] The recurrence of the dream expresses the persistent attempt of the unconscious to bring the dream content before the conscious mind.

of human use, of whatever size. The next moment each of these chance forms vanished without trace. Here is another dream of the same type: *The dreamer was walking in a wooded mountain region. At the top of a steep slope he came to a ridge of rock honeycombed with holes, and there he found a little brown man of the same colour as the iron oxide with which the rock was coated.*[73] *The little man was busily engaged in hollowing out a cave, at the back of which a cluster of columns could be seen in the living rock. On the top of each column was a dark brown human head with large eyes, carved with great care out of some very hard stone, like lignite. The little man freed this formation from the amorphous conglomerate surrounding it. The dreamer could hardly believe his eyes at first, but then had to admit that the columns were continued far back into the living rock and must therefore have come into existence without the help of man. He reflected that the rock was at least half a million years old and that the artefact could not possibly have been made by human hands.*[74]

These dreams seem to point to the presence of a formal factor in nature. They describe not just a *lusus naturae,* but the meaningful coincidence of an absolutely natural product with a human idea apparently independent of it. This is what the dreams are obviously saying,[75] and

[73] An Anthroparion or "metallic man."

[74] Cf. Kepler's theories quoted above.

[75] Those who find the dreams unintelligible will probably suspect them of harbouring quite a different meaning which is more in accord with their preconceived opinions. One can indulge in wishful thinking about dreams just as one can about anything else. For my part I prefer to keep as close to the dream statement

what they are trying to bring nearer to consciousness through repetition.

---

as possible, and to try to formulate it in accordance with its manifest meaning. If it proves impossible to relate this meaning to the conscious situation of the dreamer, then I frankly admit that I do not understand the dream, but I take good care not to juggle it into line with some preconceived theory.

CHAPTER FOUR

CONCLUSION

I do not regard these statements as in any way a final proof of my views, but simply as a conclusion from empirical premises which I would like to submit to the consideration of my reader. From the material before us I can derive no other hypothesis that would adequately explain the facts (including the ESP experiments). I am only too conscious that synchronicity is a highly abstract and "irrepresentable" quantity. It ascribes to the moving body a certain psychoid property which, like space, time, and causality, forms a criterion of its behaviour. We must completely give up the idea of the psyche's being somehow connected with the brain, and remember instead the "meaningful" or "intelligent" behaviour of the lower organisms, which are without a brain. Here we find ourselves much closer to the formal factor which, as I have said, has nothing to do with brain activity.

If that is so, then we must ask ourselves whether the relation of soul and body can be considered from this angle, that is to say whether the co-ordination of psychic and physical processes in a living organism can be understood as a synchronistic phenomenon rather than as a causal relation. Both Geulincx and Leibnitz regarded the co-ordination of the psychic and the physical as an act of God, of some principle standing outside empirical nature. The assumption of a causal relation between psyche and physis leads on the other hand to conclusions which it is difficult to square with experience: either there are physical processes which cause psychic happenings, or there is a pre-existent psyche which organizes matter. In the first case it is hard to see how chemical processes can ever produce psychic processes, and in the second case one wonders how an immaterial psyche could ever set matter in motion. It is not necessary to think of Leibnitz's pre-established harmony or anything of that kind, which would have to be absolute and would manifest itself in a universal correspondence and sympathy, rather like the meaningful coincidence of time-points lying on the same degree of latitude in Schopenhauer. The synchronicity principle possesses properties that may help to clear up the body-soul problem. Above all it is the fact of causeless order, or rather, of meaningful orderedness, that may throw light on psychophysical parallelism. The "absolute knowledge" which is characteristic of synchronistic phenomena, a knowledge not mediated by the sense organs, supports the hypothesis of a self-subsistent meaning, or even expresses its existence. Such a form of existence can only be transcendental, since, as the knowledge of future or spatially distant events shows, it is contained in a psychically relative space and time, that is to say in an irrepresentable space-time continuum.

It may be worth our while to examine more closely, from this point of view, certain experiences which seem to indicate the existence of psychic processes in what are commonly held to be unconscious states. Here I am thinking chiefly of the remarkable observations made during deep syncopes resulting from acute brain injuries. Contrary to all expectations, a severe head injury is not always followed by a corresponding loss of consciousness. To the observer, the wounded man seems apathetic, "in a trance," and not conscious of anything. Subjectively, however, consciousness is by no means extinguished. Sensory communication with the outside world is in a large measure restricted, but is not always completely cut off, although the noise of battle, for instance, may suddenly give way to a "solemn" silence. In this state there is sometimes a very distinct and impressive feeling or hallucination of levitation, the wounded man seeming to rise into the air in the same position he was in at the moment he was wounded. If he was wounded standing up, he rises in a standing position, if lying down, he rises in a lying position, if sitting, he rises in a sitting position. Occasionally his surroundings seem to rise with him—for instance the whole bunker in which he finds himself at the moment. The height of the levitation may be anything from eighteen inches to several yards. All feeling of weight is lost. In a few cases the wounded think they are making swimming movements with their arms. If there is any perception of their surroundings at all, it seems to be mostly imaginary, i.e., composed of memory images. During levitation the mood is predominantly euphoric. " 'Buoyant, solemn, heavenly, serene, relaxed, blissful, expectant, exciting' are the words used to describe it. . . . There are various kinds of 'ascension experiences.' "[1] Jantz and Beringer rightly point out that the wounded can be roused from their syncope by remarkably small stimuli, for instance if they are addressed by name or touched, whereas the most terrific bombardment has no effect.

Much the same thing can be observed in deep comas resulting from other causes. I would like to give an example from my own medical experience: A woman patient, whose reliability and truthfulness I have no reason to doubt, told me that her first birth was very difficult. After thirty hours of fruitless labour the doctor considered that a forceps delivery was indicated. This was carried out under light narcosis. She was badly torn and suffered great loss of blood. When the doctor, her mother, and her husband had gone, and everything was cleared up, the nurse wanted to eat, and the patient saw her turn round at the door and ask, "Do you want anything before I go to supper?" She tried to answer, but couldn't. She had the feeling that she was sinking through the bed into a bottomless void. She saw the nurse hurry to the bedside and seize her hand in order to take her pulse. From the way she moved her fingers to and fro the patient thought it must be almost imperceptible. Yet she herself felt quite all right, and was slightly amused at the nurse's alarm. She was not in the least frightened. That was the last she could remember for a long time. The next thing she was aware of was that, without feeling her body and its position, she was *looking down* from a point in the

[1] Hubert Jantz and Kurt Beringer, "Das Syndrom des Schwebeerlebnisses unmittelbar nach Kopfverletzungen," *Der Nervenarzt* (Berlin), XVII (1944), 202.

ceiling and could see everything going on in the room below her: she saw herself lying in the bed, deadly pale, with closed eyes. Beside her stood the nurse. The doctor paced up and down the room excitedly, and it seemed to her that he had lost his head and didn't know what to do. Her relatives crowded to the door. Her mother and her husband came in and looked at her with frightened faces. She told herself it was too stupid of them to think she was going to die, for she would certainly come round again. All this time she knew that behind her was a glorious, park-like landscape shining in the brightest colours, and in particular an emerald green meadow with short grass, which sloped gently upwards beyond a wrought-iron gate leading into the park. It was spring, and little gay flowers such as she had never seen before were scattered about in the grass. The whole demesne sparkled in the sunlight, and all the colours were of an indescribable splendour. The sloping meadow was flanked on both sides by dark green trees. It gave her the impression of a clearing in the forest, never yet trodden by the foot of man. "I knew that this was the entrance to another world, and that if I turned round to gaze at the picture directly, I should feel tempted to go in at the gate, and thus step out of life." She did not actually *see* this landscape, as her back was turned to it, but she *knew* it was there. She felt there was nothing to stop her from entering in through the gate. She only knew that she would turn back to her body and would not die. That was why she found the agitation of the doctor and the distress of her relatives stupid and out of place.

The next thing that happened was that she awoke

from her coma and saw the nurse bending over her in bed. She was told that she had been unconscious for about half an hour. The next day, some fifteen hours later, when she felt a little stronger, she made a remark to the nurse about the incompetent and "hysterical" behaviour of the doctor during her coma. The nurse energetically denied this criticism in the belief that the patient had been completely unconscious at the time and could therefore have known nothing of the scene. Only when she described in full detail what had happened during the coma was the nurse obliged to admit that the patient had perceived the events exactly as they happened in reality.

One might conjecture that this was simply a psychogenic twilight state in which a split-off part of consciousness still continued to function. The patient, however, had never been hysterical and had suffered a genuine heart collapse followed by syncope due to cerebral anemia, as all the outward and evidently alarming symptoms indicated. She really was in a coma and ought to have had a complete psychic black-out and been altogether incapable of clear observation and sound judgment. The remarkable thing was that it was not an immediate perception of the situation through indirect or unconscious observation, but she saw the whole situation from *above,* as though "her eyes were in the ceiling," as she put it.

Indeed, it is not easy to explain how such unusually intense psychic processes can take place, and be remembered, in a state of severe collapse, and how the patient could observe actual events in concrete detail with closed eyes. One would expect such obvious cerebral anemia to militate against or prevent the occurrence of highly complex psychic processes of that kind.

Sir Auckland Geddes presented a very similar case before the Royal Medical Society on February 26, 1927, though here the ESP went very much further. During a state of collapse the patient noted the splitting off of an integral consciousness from his bodily consciousness, the latter gradually resolving itself into its organ components. The other consciousness possessed verifiable ESP.[2]

These experiences seem to show that in swoon states, where by all human standards there is every guarantee that conscious activity and sense perception are suspended, consciousness, reproducible ideas, acts of judgment, and perceptions can still continue to exist. The accompanying feeling of levitation, alteration of the angle of vision, and extinction of hearing and of coenaesthetic perceptions indicate a shift in the localization of consciousness, a sort of separation from the body, or from the cerebral cortex or cerebrum which is conjectured to be the seat of conscious phenomena. If we are correct in this assumption, then we must ask ourselves whether there is some other nervous substrate in us, apart from the cerebrum, that can think and perceive, or whether the psychic processes that go on in us during loss of consciousness are synchronistic phenomena, i.e., events which have no causal connection with organic processes. This last possibility cannot be rejected out of hand in view of the existence of ESP, i.e., of perceptions independent of space and time which cannot

[2] Cf. G. N. M. Tyrrell's report in *The Personality of Man* (London, 1947), pp. 197 *f.* There is another case of this kind on pp. 199 *f.*

be explained as processes in the biological substrate. Where sense perceptions are impossible from the start, it can hardly be a question of anything but synchronicity. But where there are spatial and temporal conditions which would make perception and apperception possible in principle, and only the activity of consciousness, or the cortical function, is extinguished, and where, as in our example, a conscious phenomenon like perception and judgment nevertheless occurs, then the question of a nervous substrate might well be considered. It is well nigh axiomatic that conscious processes are tied to the cerebrum, and that the lower centres contain nothing but chains of reflexes which in themselves are unconscious. This is particularly true of the sympathetic system. Hence the insects, which have no cerebrospinal nervous system at all, but only a double chain of ganglia, are regarded as reflex automata.

This view has recently been challenged by the researches which von Frisch, of Graz, made into the life of bees. It turns out that bees not only tell their comrades, by means of a peculiar sort of dance, that they have found a feeding-place, but that they also indicate its direction and distance, thus enabling the beginners to fly to it directly.[3] This kind of message is no different in principle from information conveyed by a human being. In the latter case we would certainly regard such behaviour as a conscious and intentional act and can hardly imagine how anyone could prove in a court of law that it had taken place unconsciously. We could, at a pinch, admit on the

[3] Karl von Frisch, *The Dancing Bees,* trans. by Dora Ilse (New York and London, 1954), pp. 112 *ff.*

basis of psychiatric experiences that objective information can in exceptional cases be communicated in a twilight state, but would expressly deny that communications of this kind are normally unconscious. Nevertheless it would be possible to suppose that in bees the process is unconscious. But that would not help to solve the problem, because we are still faced with the fact that the ganglionic system apparently achieves exactly the same result as our cerebral cortex. Nor is there any proof that bees are unconscious.

Thus we are driven to the conclusion that a nervous substrate like the sympathetic system, which is absolutely different from the cerebrospinal system in point of origin and function, can evidently produce thoughts and perceptions just as easily as the latter. What then are we to think of the sympathetic system in vertebrates? Can it also produce or transmit specifically psychic processes? Von Frisch's observations prove the existence of transcerebral thought and perception. One must bear this possibility in mind if we want to account for the existence of some form of consciousness during an unconscious coma. During a coma the sympathetic system is not paralysed and could therefore be considered as a possible carrier of psychic functions. If that is so, then one must ask whether the normal state of unconsciousness in sleep, and the potentially conscious dreams it contains, can be regarded in the same light—whether, in other words, dreams are produced not so much by the activity of the sleeping cortex, as by the unsleeping sympathetic system, and are therefore of a transcerebral nature.

Outside the realm of psychophysical parallelism,

which we cannot at present pretend to understand, synchronicity is not a phenomenon whose regularity it is at all easy to demonstrate. One is as much impressed by the disharmony of things as one is surprised by their occasional harmony. In contrast to the idea of a pre-established harmony, the synchronistic factor merely claims the existence of an intellectually necessary principle which could be added as a fourth to the recognized triad of space, time, and causality. These factors are necessary but not absolute—most psychic contents are non-spatial, time and causality are psychically relative—and in the same way the synchronistic factor proves to be only conditionally valid. But unlike causality, which reigns despotically over the whole picture of the macrophysical world and whose universal rule is shattered only in certain lower orders of magnitude, synchronicity is a phenomenon that seems to be primarily connected with psychic conditions, that is to say with processes in the unconscious. Synchronistic phenomena are found to occur—experimentally—with some degree of regularity and frequency in the intuitive, "magical" procedures, where they are subjectively convincing but are extremely difficult to verify objectively and cannot be statistically evaluated (at least at present).

On the organic level it might be possible to regard biological morphogenesis in the light of the synchronistic factor. Professor A. M. Dalcq (of Brussels) understands form, despite its tie with matter, as a "continuity that is superordinate to the living organism." [4] Sir James Jeans

[4] "La Morphogénèse dans la cadre de la biologie générale," *Verhandlungen der Schweizerischen Naturforschenden Gesellschaft* (129th Annual Meeting, at Lausanne; pub. Aarau, 1949), 37-72.

reckons radioactive decay among the causeless events which, as we have seen, include synchronicity. He says: "Radioactive break-up appeared to be an effect without a cause, and suggested that the ultimate laws of nature were not even causal." [5] This highly paradoxical formula, coming from the pen of a physicist, is typical of the intellectual dilemma with which radioactive decay confronts us. It, or rather the phenomenon of "half-life," appears as an instance of acausal orderedness—a conception which also includes synchronicity and to which I shall revert below.

Synchronicity is not a philosophical view but an empirical concept which postulates an intellectually necessary principle. This cannot be called either materialism or metaphysics. No serious investigator would assert that the nature of what is observed to exist, and of that which observes, namely the psyche, are known and recognized quantities. If the latest conclusions of science are coming nearer and nearer to a unitary idea of being, characterized by space and time on the one hand and by causality and synchronicity on the other, that has nothing to do with materialism. Rather it seems to show that there is some possibility of getting rid of the incommensurability between the observed and the observer. The result, in that case, would be a unity of being which would have to be expressed in terms of a new conceptual language—a "neutral language," as W. Pauli once called it.

Space, time, and causality, the triad of classical phys-

Cf. above, the similar conclusion reached by the zoologist A. C. Hardy.

[5] *Physics and Philosophy* (Cambridge, 1942), p. 127; cf. also p. 151.

ics, would then be supplemented by the synchronicity factor and become a tetrad, a *quaternio* which makes possible a whole judgment:

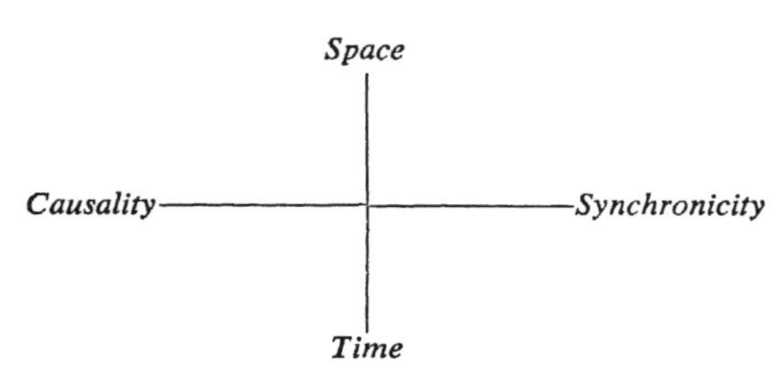

Here synchronicity is to the three other principles as the one-dimensionality of time [6] is to the three-dimensionality of space, or as the recalcitrant "Fourth" in the *Timaeus,* which, Plato says, can only be added "by force" to the other three.[7] Just as the introduction of time as the fourth dimension in modern physics postulates an irrepresentable space-time continuum, so the idea of synchronicity with its inherent quality of meaning produces a picture of the world so irrepresentable as to be quite baffling.[8] The advantage, however, of adding this concept is that it makes possible a view which includes the psychoid factor in our description and knowledge of nature—an

[6] I am not counting P. A. M. Dirac's multi-dimensionality of time.

[7] Cf. my "Versuch einer psychologischen Deutung des Trinitätsdogmas," in *Symbolik des Geistes* (Zurich, 1948), pp. 323 *ff.*

[8] Sir James Jeans (*Physics and Philosophy,* p. 215) thinks it possible "that the springs of events in this substratum include our own mental activities, so that the future course of events may depend in part on these mental activities." The causalism of this argument does not seem to me altogether tenable.

a priori meaning or "equivalence." The problem that runs like a red thread through the speculations of alchemists for fifteen hundred years thus repeats and solves itself, the so-called axiom of Maria the Jewess (or Copt): "Out of the Third comes the One as the Fourth." [9] This cryptic observation confirms what I said above, that in principle new points of view are not as a rule discovered in territory that is already known, but in out-of-the-way places that may even be avoided because of their bad name. The old dream of the alchemists, the transmutation of chemical elements, this much-derided idea, has become a reality in our own day, and its symbolism, which was no less an object of ridicule, has turned out to be a veritable goldmine for the psychology of the unconscious. Their dilemma of three and four, which began with the story that serves as a setting for the *Timaeus* and extends all the way to the Cabiri scene in *Faust,* Part II, is recognized by a sixteenth-century alchemist, Gerhard Dorn, as the decision between the Christian Trinity and the *serpens quadricornutus,* the four-horned serpent who is the Devil. As though in anticipation of things to come he anathematizes the pagan quaternity which was ordinarily so beloved of the alchemists, on the ground that it arose from the binarius (the number 2) and is thus something material, feminine, and devilish.[10] Dr. von Franz has demonstrated this emergence of trinitarian thinking in the *Parable* of Bernard of Treviso, in Khunrath's *Amphitheatrum,* in Michael Maier, and in the

[9] "ἐκ τοῦ τρίτου τὸ ἓν τέταρτον." Cf. *Psychology and Alchemy* (New York and London, 1953), p. 23.

[10] "De tenebris contra naturam," in *Theatrum chemicum* (Ursel, 1602), I, pp. 540 *ff.*

anonymous author of the *Aquarium sapientum*.[11] W. Pauli calls attention to the polemical writings of Kepler and of Robert Fludd, in which Fludd's correspondence theory was the loser and had to make room for Kepler's theory of three principles.[12] The decision in favour of freedom, which in certain respects ran counter to the alchemical tradition, was followed by a scientific epoch that knew nothing of correspondence and clung with passionate insistence to a triadic view of the world—a continuation of the trinitarian type of thinking—which described and explained everything in terms of space, time, and causality.

The revolution brought about by the discovery of radioactivity has considerably modified the classical views of physics. So great is the change of standpoint that we have to revise the classical schema I made use of above. As I was able, thanks to the friendly interest which Professor Pauli evinced in my work, to discuss these questions of principle with a professional physicist who could at the same time appreciate my psychological arguments, I am in a position to put forward a suggestion that takes modern physics into account. Pauli suggested replacing the opposition of space and time in the classical schema by (conservation of) energy and the space-time continuum. This suggestion led me to a closer definition of the other pair of opposites—causality and synchronicity—with a view to establishing some kind of connection between

[11] Marie-Louise von Franz, "Die Parabel von der Fontina des Grafen von Tarvis" (unpublished MS.).

[12] See Pauli's contribution to the present volume.

these two heterogeneous concepts. We finally agreed on the following *quaternio:*

Indestructible Energy

Constant Connection through Effect (Causality)

Inconstant Connection through Contingency, Equivalence, or "Meaning" (Synchronicity)

Space-Time Continuum

This schema satisfies on the one hand the postulates of modern physics, and on the other hand those of psychology. The psychological point of view needs clarifying. A causalistic explanation of synchronicity seems out of the question for the reasons given above. It consists essentially of "chance" equivalences. Its *tertium comparationis* rests on the psychoid factors I call archetypes. These are *indefinite,* that is to say they can be known and determined only approximately. Although associated with causal processes, or "carried" by them, they continually go beyond their frame of reference, an infringement to which I would give the name "transgressivity," because the archetypes are not found exclusively in the psychic sphere, but can occur just as much in circumstances that are not psychic (equivalence of an outward physical process with a psychic one). Archetypal equivalences are *contingent* to causal determination, that is to say there exist between them and the causal processes no relations that conform to law. They seem, therefore, to represent a special instance of randomness or chance, or of that "random state" which "runs through time in a way that fully conforms to law," as Andreas Speiser says.[13] It is an initial state which is "not governed by mechanistic law" but is the precondition of law, the chance substrate on which law is based. If we consider synchronicity or the archetypes as the contingent, then the latter takes on the specific aspect of a modality that has the functional significance of a world-constituting factor. The archetype represents *psychic probability,* since it portrays ordinary instinctual events in the form of *types*. It is a special psychic instance of probability in general, which "is made up of the laws of chance and lays down rules for nature just as the laws of mechanics do." [14] We must agree with Speiser that although in the realm of pure intellect the contingent is "a formless substance," it reveals itself to psychic introspection—so far as inward perception can grasp it at all—as an image, or rather a type which underlies not only the psychic equivalences but, remarkably enough, the psychophysical equivalences too.

It is difficult to divest conceptual language of its causalistic colouring. Thus the word "underlying," despite its causalistic connotation, does not refer to anything causal, but simply to an existing quality, an irreducible contingency which is "Just-So." The meaningful coincidence or equivalence of a psychic and a physical state that have no causal relationship to one another means, in general terms, that it is a modality without a cause, an "acausal orderedness." The question now arises whether our definition of synchronicity with reference to the equiv-

[13] Über die Freiheit," *Basler Universitätsreden,* XXVIII (1950), 4 *f.*

[14] Ibid., p. 6.

alence of psychic and physical processes is capable of expansion, or rather, requires expansion. This requirement seems to force itself on us when we consider the above, wider conception of synchronicity as an "acausal orderedness." Into this category come all "acts of creation," *a priori* factors such as the properties of natural numbers, the discontinuities of modern physics, etc. Consequently we would have to include constant and experimentally reproducible phenomena within the scope of our expanded concept, though this does not seem to accord with the nature of the phenomena included in synchronicity narrowly understood. The latter are mostly individual cases which cannot be repeated experimentally. This is not of course altogether true, as Rhine's experiments show and numerous other experiences with clairvoyant individuals. These facts prove that even in individual cases which have no common measure there are certain regularities and therefore constant factors, from which we must conclude that our narrower conception of synchronicity is probably too narrow and really needs expanding. I incline in fact to the view that synchronicity in the narrower sense is only a particular instance of general acausal orderedness—that, namely, of the equivalence of psychic and physical processes where the observer is in the fortunate position of being able to recognize the *tertium comparationis*. But as soon as he perceives the archetypal background he is tempted to trace the mutual assimilation of independent psychic and physical processes back to a (causal) effect of the archetype, and thus to overlook the fact that they are merely contingent. This danger is avoided if one regards synchronicity as a special instance of general acausal orderedness. In this way we also avoid multiplying our principles of explanation illegitimately, for the archetype *is* the introspectively recognizable form of *a priori* psychic orderedness. If an external synchronistic process now associates itself with it, it falls into the same basic pattern—in other words, it too is "ordered." This form of orderedness differs from that of the properties of natural numbers or the discontinuities of physics in that the latter have existed from eternity and occur regularly, whereas the forms of psychic orderedness are *acts of creation in time*. That, incidentally, is precisely why I have stressed the element of time as being characteristic of these phenomena and called them *synchronistic*.

The modern discovery of discontinuity (e.g., the orderedness of energy quanta, of radium decay, etc.) has put an end to the sovereign rule of causality and thus to the triad of principles. The territory lost by the latter belonged earlier to the sphere of correspondence and sympathy, concepts which reached their greatest development in Leibnitz's idea of pre-established harmony. Schopenhauer knew far too little about the empirical foundations of correspondence to realize how hopeless his causalistic attempt at explanation was. Today, thanks to the ESP experiments, we have a great deal of empirical material at our disposal. We can form some conception of its reliability when we learn from G. E. Hutchinson [15] that the ESP experiments conducted by S. G. Soal and K. M. Goldney have a probability of $1 : 10^{35}$, this being equivalent to the number of molecules in 250,000 tons of water. There are relatively few experi-

[15] S. G. Soal, "Science and Telepathy," *Enquiry* (London), I (1948): 2, p. 6.

ments in the field of the natural sciences whose results come anywhere near so high a degree of certainty. The exaggerated scepticism in regard to ESP is really without a shred of justification. The main reason for it is simply the ignorance which nowadays, unfortunately, seems to be the inevitable accompaniment of specialism and screens off the necessarily limited horizon of specialist studies from all higher and wider points of view in the most undesirable way. How often have we not found that the so-called "superstitions" contain a core of truth that is well worth knowing! It may well be that the originally magical significance of the word "wish," which is still preserved in "wishing-rod" (divining rod, or magic wand) and expresses not just wishing in the sense of desire but a magical action,[16] and the traditional belief in the efficacy of prayer, are both based on the experience of concomitant synchronistic phenomena.

Synchronicity is no more baffling or mysterious than the discontinuities of physics. It is only the ingrained belief in the sovereign power of causality that creates intellectual difficulties and makes it appear unthinkable that causeless events exist or could ever occur. But if they do, then we

[16] Jacob Grimm, *Teutonic Mythology,* trans. by J. S. Stallybrass (London, 1883–88), I, p. 137. Wish-objects are magic implements forged by dwarfs, such as Odin's spear Gungnir, Thor's hammer Mjollnir, and Freya's sword (II, p. 870). Wishing is "gotes kraft" (divine power). "Got hât an sie den wunsch geleit und der wünschelruoten hort" (God has bestowed the wish on her and the treasure of [*or:* found by] the wishing-rod). "Beschoenen mit wunsches gewalte" (to make beautiful with the power of the wish) (IV, p. 1329). "Wish" = Sanskrit *manoratha,* literally, "car of the mind" or of the psyche, i.e., wish, desire, fancy. (A. A. Macdonell, *A Practical Sanskrit Dictionary,* London, 1924, s.v.)

must regard them as *creative acts,* as the continuous creation [17] of a pattern that exists from all eternity, repeats itself sporadically, and is not derivable from any known antecedents. We must of course guard against thinking of every event whose cause is unknown as "causeless." This, as I have already stressed, is admissible only when a cause is not even thinkable. But thinkability is itself an idea that needs the most rigorous criticism. Had the atom [18] corresponded to the original philosophical conception of it, its fissionability would be unthinkable. But once it proves to be a measurable quantity, its non-fissionability becomes unthinkable. Meaningful coincidences are thinkable as

[17] Continuous creation is to be thought of not only as a series of successive acts of creation, but also as the eternal presence of the *one* creative act, in the sense that God "was always the Father and always generated the Son" (Origen, *De principiis,* I, 2, 3), or that he is the "eternal Creator of minds" (Augustine, *Confessions,* XI, 31, trans. F. J. Sheed, London, 1943, p. 273). God is contained in his own creation, "nor does he stand in need of his own works, as if he had place in them where he might abide; but endures in his own eternity, where he abides and creates whatever pleases him, both in heaven and earth" (Augustine, on Ps. 113 : 14, in *Expositions on the Book of Psalms,* Library of Fathers of the Holy Catholic Church, Vol. V, Oxford, 1853). What happens successively in time is simultaneous in the mind of God: "An immutable order binds mutable things into a pattern, and in this order things which are not simultaneous in time exist simultaneously outside time" (Prosper of Aquitaine, *Sententiae ex Augustino delibatae,* XLI [Migne, *P.L.,* LI, col. 433]). "Temporal succession is without time in the eternal wisdom of God" (LVII [Migne, col. 455]). Before the Creation there was no time—time only began with created things: "Rather did time arise from the created than the created from time" (CCLXXX [Migne, col. 468]). "There was no time before time, but time was created together with the world" (Anon., *De triplici habitaculo,* VI [Migne, *P.L.,* XL, col. 995]).

[18] [From ἄτομος, "indivisible, that cannot be cut."—Trans.]

pure chance. But the more they multiply and the greater and more exact the correspondence is, the more their probability sinks and their unthinkability increases, until they can no longer be regarded as pure chance but, for lack of a causal explanation, have to be thought of as meaningful arrangements. As I have already said, however, their "inexplicability" is not due to the fact that the cause is unknown, but to the fact that a cause is not even thinkable in intellectual terms. This is necessarily the case when space and time lose their meaning or have become relative, for under those circumstances a causality which presupposes space and time for its continuance can no longer be said to exist and becomes altogether unthinkable.

For these reasons it seems to me necessary to introduce, alongside space, time, and causality, a category which not only enables us to understand synchronistic phenomena as a special class of natural events, but also takes the contingent partly as a universal factor existing from all eternity, and partly as the sum of countless individual acts of creation occurring in time.

RÉSUMÉ

I have been informed that many readers find it difficult to follow my argument. Acausality and the idea of synchronicity as such, and also the astrological experiment, seem to present especial difficulties to their understanding, and for this reason I should like to make a few additional remarks in order to sum up these three points.

1. ACAUSALITY. If natural law were an absolute truth, then of course there could not possibly be any processes that deviate from it. But since causality is a *statistical* truth, it holds good only on average and thus leaves room for *exceptions* which must somehow be experienceable, that is to say, *real.* I try to regard synchronistic events as acausal exceptions of this kind. They prove to be relatively independent of space and time; they relativize space and time in so far as space presents in principle no obstacle to their passage and the sequence of events in time is inverted, so that it looks as if an event which has not yet occurred were causing a perception in the present. But if space and time are relative, then causality too loses its validity, since the sequence of cause and effect is either relativized or abolished.

2. SYNCHRONICITY. Despite my express warning I see that this concept has already been confused by the critics with *synchronism.* By synchronicity I mean the occurrence of a *meaningful coincidence in time.* It can take three forms:

a) The coincidence of a certain psychic content with a corresponding objective process which is perceived to take place simultaneously.

b) The coincidence of a subjective psychic state with a phantasm (dream or vision) which later turns out to be a more or less faithful reflection of a "synchronistic," objective event that took place more or less simultaneously, but at a distance.

c) The same, except that the event perceived takes place in the future and is represented in the present only by a phantasm that corresponds to it.

Whereas in the first case an objective event coincides with a subjective content, the synchronicity in the other two cases can only be verified subsequently, though the synchronistic event as such is formed by the coincidence of a neutral psychic state with a phantasm (dream or vision).

3. THE ASTROLOGICAL EXPERIMENT. The fifty possible aspects to be considered in a marriage relationship were examined statistically. The result showed that in three quite fortuitously assorted batches of marriage horoscopes the greatest frequency fell to three different lunar conjunctions. The probability of these three figures is far from being significant, because, if we take any number of marriage horoscopes, there is always a 1 : 1500 probability of our getting a similar result. Even so, this probability is not so great that we could expect a repetition of the same result in the second batch, for the probability in this case already amounts to $1 : 1500^2$. This figure is so high that a repetition of the first result must be regarded as extremely improbable. If it is now repeated for

a third time, there is sufficient ground for assuming a synchronistic phenomenon. The probability here amounts to 1 : 2,500,000. In addition, the lunar conjunctions I have mentioned are not just *any* conjunctions, but correspond to the three main pillars of the horoscope and appear in that order: ☾ ☌ ☉, ☾ ☌ ☾, ☾ ☌ *Asc*. Furthermore, these lunar conjunctions are the classical aspects for marriage.

This meaningful arrangement is an exceedingly improbable one. It looks very like a conscious fraud, and its calculation was bedevilled by all manner of unconscious tendencies to twist the result in favour of astrology and synchronicity, despite there being a contrary tendency in the conscious minds of the persons responsible.

W. PAULI

The Influence of Archetypal Ideas on the Scientific Theories of Kepler

Translated from the German by

PRISCILLA SILZ

PREFATORY NOTE

It is my pleasant duty to express my warmest thanks to all those who have given me assistance and encouragement in the writing and publication of this essay.

In particular I owe a debt of gratitude to Professor Erwin Panofsky, of the Institute for Advanced Study, at Princeton, for many discussions of the problems here concerned in the light of the history of ideas, also for his procurement and critical appraisal of original texts and for several translations from the Latin; to Professor C. G. Jung and Dr. C. A. Meier for detailed and essential discussions connected with the psychological aspect of the formation of scientific concepts and their archetypal basis; and to Dr. M.-L. von Franz for her translations from the Latin, the most numerous and most important in the essay, as well as for a painstaking and often wearisome examination of different original texts. In the English edition, her translations have been revised by Professor Panofsky. I may add that the English edition embodies a few minor corrections.

W. PAULI

THE INFLUENCE OF ARCHE-TYPAL IDEAS ON THE SCIENTIFIC THEORIES OF KEPLER

1

Although the subject of this study is an historical one, its purpose is not merely to enumerate facts concerning scientific history or even primarily to present an appraisal of a great scientist, but rather to illustrate particular views on the origin and development of concepts and theories of natural science in the light of one historic example. In so doing we shall also have occasion to discuss the significance for modern science of the problems which arose in the period under consideration, the seventeenth century.

In contrast to the purely empirical conception according to which natural laws can with virtual certainty, be derived from the material of experience alone, many physicists have recently emphasized anew the fact that intuition and the direction of attention play a considerable role in the development of the concepts and ideas, generally far transcending mere experience, that are necessary for the erection of a system of natural laws (that is, a scientific theory). From the standpoint of this not purely empiristic conception, which we also accept, there arises the question, What is the nature of the bridge between the sense perceptions and the concepts? All logical thinkers have arrived at the conclusion that pure logic is fundamentally incapable of constructing such a link. It seems most satisfactory to introduce at this point the postulate of a cosmic order independent of our choice and distinct from the world of phenomena. Whether one speaks of the "participation of natural things in ideas" or of a "behaviour of metaphysical things—those, that is, which are in themselves real," the relation between sense perception and idea remains predicated upon the fact that both the soul of the perceiver and that which is recognized by perception are subject to an order thought to be objective.

Every partial recognition of this order in nature leads to the formulation of statements that, on the one hand, concern the world of phenomena and, on the other, transcend it by employing, "idealizingly," general logical concepts. The process of understanding nature as well as the happiness that man feels in understanding, that is, in the conscious realization of new knowledge, seems thus to be based on a correspondence, a "matching" of inner images pre-existent in the human psyche with external objects and their behaviour. This interpretation of scientific knowledge, of course, goes back to Plato and is, as we shall see, very clearly advocated by Kepler. He speaks in fact of ideas that are pre-existent in the mind of God and were implanted in the soul, the image of God, at the time of creation. These primary images which the soul can perceive with the aid of an innate "instinct" are called by Kepler archetypal ("archetypalis"). Their agreement with the "primordial images" or *archetypes* introduced into modern psychology by C. G. Jung and functioning as "instincts of imagination" is very extensive. When modern psychology brings proof to show that all understanding is a long-drawn-out process initiated by processes in the unconscious long before the content of consciousness can be rationally formulated, it has directed attention again to the preconscious, archaic level of cognition. On this level the place of clear concepts is taken by images with strong emotional content, not thought out but beheld, as it were, while being painted. Inasmuch as these images are an "expression of a dimly suspected but still unknown state of affairs" they can also be termed symbolical, in accordance with the definition of the symbol proposed by C. G. Jung. As *ordering* operators and image-formers in this world of symbolical images, the archetypes thus function as the sought-for bridge between the sense perceptions and the ideas and are, accordingly, a necessary presupposition even for evolving a scientific theory of nature. However, one must guard against transferring this *a priori* of knowledge into the conscious mind and relating it to definite ideas capable of rational formulation.

2

As a consequence of the rationalistic attitude of scientists since the eighteenth century, the background processes that accompany the development of the natural sciences, although present as always and of decisive effect, remained to a large extent unheeded, that is to say, confined to the unconscious. On the other hand, in the Middle Ages down to the beginning of modern times, we have no natural science in the present-day sense but merely that pre-scientific stage, just mentioned, of a magical-symbolical description of nature. This, of course, is also to be found in alchemy, the psychological significance of which has been the subject of intensive investigation by C. G. Jung. My attention was therefore directed especially to the seventeenth century, when, as the fruit of a great intellectual effort, a truly scientific way of thinking, quite new at the time, grew out of the nourishing soil of a magical-animistic conception of nature. For the purpose of illustrating the relationship between archetypal ideas and scientific theories of nature Johannes Kepler (1571–1630) seemed to me especially suitable, since his ideas represent a remarkable intermediary stage between the earlier, magical-symbolical and the modern, quantitive-mathematical descriptions of nature.[1]

[1] The chief writings of Kepler are:

Mysterium cosmographicum (Tübingen, 1st edn., 1596; 2nd edn., 1621).

Ad Vitellionem paralipomena, quibus astronomiae pars optica traditur (Frankfort on the Main, 1604).

De stella nova in pede serpentarii (Prague, 1606).

De motibus stellae Martis (Prague, 1609).

Tertius interveniens (Frankfort on the Main, 1610).

Dioptrice (Augsburg, 1611).

Harmonices mundi (in five books, Augsburg, 1619).

Epitome astronomiae Copernicanae (Linz and Frankfort on the Main, 1618–21).

It should be noted that Newton's chief work, *Philosophiae naturalis Principia mathematica,* appeared in 1687.

In that age many things that, later on, were to be divided by a critical effort were still closely interrelated: the view of the universe was not as yet split into a religious one and a scientific one. Religious meditations, an almost mathematical symbol of the Trinity, modern optical theorems, essential discoveries in the theory of vision and the physiology of the eye (such as the proof that the retina is the sensitive organ of the eye), are all to be found in the same book, *Ad Vitellionem paralipomena*. Kepler is a passionate adherent of the Copernican heliocentric system, on which he wrote the first coherent textbook (*Epitome astronomiae Copernicanae*). The connection of his heliocentric creed—as I should like to call it, in intentional allusion to religious creeds—with the particular form of his Protestant-Christian religion in general and with his archetypal ideas and symbols in particular will be examined in detail on the following pages.

On the basis of the heliocentric conception, Kepler discovered his three famous laws of planetary motion: 1. Revolution in ellipses, the sun being located in one of the foci, in *De motibus stellae Martis*. 2. Radius vector of each planet covering equal areas in equal time, in *De motibus stellae Martis*. 3. Time of revolution τ proportional to $a^{3/2}$, a being half the major axis, in *Harmonices mundi,* Book V. Not very long after their discovery, these laws that today have a place in all text-books were to become one of the pillars upon which Newton [2] based his theory of gravitation, namely, the law of the diminution of the force of gravitation in inverse proportion to

[2] *Principia.*

the square of the distance of the heavy masses from each other.

But these laws that Kepler discovered—the third after years of effort—are not what he was originally seeking. He was fascinated by the old Pythagorean idea of the music of the spheres (which, incidentally, also played no small part in contemporary alchemy) and was trying to find in the movement of the planets the same proportions that appear in the harmonious sounds of tones and in the regular polyhedra. For him, a true spiritual descendant of the Pythagoreans, all beauty lies in the correct proportion, for "Geometria est archetypus pulchritudinis mundi" (Geometry is the archetype of the beauty of the world). This axiom of his is at once his strength and his limitation: his ideas on regular bodies and harmonious proportions did, after all, not quite work out in the planetary system, and a trend of research, like that of his contemporary Galileo, which directed attention to the constant acceleration of freely falling bodies, was quite foreign to Kepler's attitude, since for such a trend the de-animation of the physical world, which was to be completed only with Newton's *Principia,* had not as yet progressed far enough. In Kepler's view the planets are still living entities endowed with individual souls. Since the earth had lost its unique position among the planets he had to postulate also an *anima terrae.* We shall see how the souls of the heavenly bodies play an essential role in Kepler's particular views on astrology. Yet the de-animation of the physical world had already begun to operate in Kepler's thought. He does, to be sure, occasionally mention the alchemical world-soul, the *anima mundi,* that sleeps in matter, is made responsible for the origin of a new star (*De stella nova,* Ch. 24), and is said to have its seat—that is, its special concentration—in the sun. But it can be seen clearly that the *anima mundi* is no more than a kind of relic in Kepler's mind and plays a subordinate role compared to the individual souls of the various heavenly bodies. Although Kepler's ideas reveal quite unmistakably the influence of Paracelsus and his pupils, the contrast between his scientific method of approach and the magical-symbolical attitude of alchemy was nevertheless so strong that Fludd, in his day a famous alchemist and Rosicrucian, composed a violent polemic against Kepler's chief work, *Harmonices mundi.* In Section 6 we shall revert to this polemic, in which two opposing intellectual worlds collided.

Before I go into detail regarding Kepler's ideas, I shall furnish some brief biographical data to illuminate the historical background of his life. Kepler was born in 1571 in the town of Weil in Württemberg. He was brought up in the Protestant faith; indeed, his parents originally intended him for the clergy. But early, because of his profession of the Copernican doctrine, he came into conflict with the Evangelical theology prevailing in Württemberg. Mästlin, his teacher in mathematics and astronomy, obtained a teaching post for him in Graz. When Kepler sent Mästlin his first work, the *Mysterium cosmographicum,* so that it might be published in Tübingen, Mästlin encountered a difficulty: the Senate objected because the doctrine of the movement of the earth, which underlay the work, might tend to diminish the prestige of Holy Writ. The objections were eventually overcome, however, and the work appeared. But later Kepler was faced with new difficulties: Archduke Ferdinand, the ruler of Styria, rigorously carried out the Counter Reformation in his lands. As a Protestant Kepler was banished from the country on pain of death. This brought him, fortunately, into contact with Tycho Brahe. In 1599, some years after the death of his patron, Frederick II of Denmark, Brahe had accepted the call of Emperor Rudolf II and moved to Prague, leaving his famous observatory Stjarneborg, on the island of Hven. From Prague in the same year came Tycho's invitation to Kepler just as the latter was being exiled from Graz. The collaboration of the two astronomers was exceedingly fruitful. Tycho, to be sure, died only two years later, but Kepler was able to derive his first two laws from Tycho's exact observations. Circles were replaced by ellipses (1609), a great revolution in astronomy!

After the death of Rudolf II, Kepler moved to Linz. He was obliged to expend great energy in the defence of his mother, who had been put to trial for witchcraft because a neighbour woman had fallen ill and maintained that she had put her under a spell. Kepler finally succeeded in rescuing his mother from torture and the stake. In 1619, the year in which Kepler published his chief work, *Harmonices mundi,* Ferdinand II, the former Archduke, ascended the imperial throne. The persecutions of the Protestants increased, and in 1626 Kepler had to relinquish his post in Linz. After negotiations he undertook with Wallenstein were broken off by the latter's downfall, Kepler travelled in 1630 to Regensburg in order to prefer his financial claims at the Imperial Diet. His health had already been weakened, and he succumbed to the trials and excitements of the journey soon after his arrival in Regensburg. He was buried outside the gates of that city; the Thirty Years War soon afterward removed every trace of his grave.

3

Kepler's grave, however, is of far less importance to us than the ideas that are clearly expressed in his well-preserved works, which we shall now examine more closely as documents of an age that despite all political and religious confusions was a period of scientific flowering.

Kepler's archetypal concepts are *hierarchically* arranged in such a way that the trinitarian Christian Godhead, incapable of visualization, occupies the highest place, and each level is an image of the one above it. As regards this, Kepler invokes the authority of the doctrine of *signatura rerum,* the signs of things, a doctrine employed and expanded by Agrippa of Nettesheim and by Paracelsus and his pupils. According to this theory, which originated in the Middle Ages and is closely connected with the old idea of the correspondence of microcosm and macrocosm, things have a hidden meaning that is expressed in their external form, inasmuch as this form points to another, not directly visible level of reality. Now for Kepler the most beautiful image, the one that represents God's own form of being (*idea ipsius essentiae*), is the three-dimensional sphere. He says already in his early work the *Mysterium cosmographicum:* "The image of the triune God is in the spherical surface, that is to say, the Father's in the centre, the Son's in the outer surface, and the Holy Ghost's in the equality of relation between point and circumference." [3] The movement or emanation passing from the center to the outer surface is for him the symbol of creation, while the curved outer surface itself is supposed to represent the eternal Being of God. Of the magnitudes (*quanta* or *quantitates*) evolved in the beginning by the Creator the curved one is the symbol of the intellectual or spiritual and is thus more perfect than the straight, which as the simulacrum of the created represents the physical world. This can be learned from the following quotation, which also shows how in Kepler's hierarchical arrangement the human mind bears the same relation to the perfect, Divine Mind as does the circle to the sphere.

. . . There follows, then, the straight line which, by the movement of a point located in the centre [of the sphere] to a single point on the surface, represents the first beginnings of creation, emulating the eternal generation of the Son in that the centre flows out towards infinitely many points of the whole surface, which, under the rule of most perfect equality, is formed and described by in-

. . . Sequitur igitur recta linea, quae ex fluxu puncti in centro in punctum unicum superficiei prima rudimenta creationis delineat, aemula aeternae generationis filii (egressu centri versus infinita puncta totius superficiei, lineis infinitis sub aequalitate omnibus perfectissima figuratae et depictae), quae recta linea elementum scilicet est formae corporeae. Haec in latum ducta jam ipsam formam corpoream

[3] *Joannis Kepleri Astronomi Opera omnia,* ed. by Ch. Frisch (8 vols., Frankfort on the Main and Erlangen, 1858 *ff.*), Vol. I, pp. 122 *f.* "Dei triuni imago in sphaerica Superficie, Patris scilicet in centro, Filii in superficie, Spiritus in aequalitate σχέσεως inter punctum et ambitum."

finitely many lines; and this straight line is, needless to say, the element of corporeal form. When widened, it [the straight line] already adumbrates the corporeal form itself, creating the plane. But when intersected by a plane, the sphere displays in this section the circle, the genuine image of the created mind, placed in command of the body which it is appointed to rule; and this circle is to the sphere as the human mind is to the Mind Divine, that is to say, as the line is to the surface; but both, to be sure, are circular. To the plane in which it is contained the circle is related as is the curve to the straight line, these two being mutually incompatible and incommensurable, and the circle beautifully fits into the intersecting plane (of which it is the circumscribing limit) as well as into the intersected sphere by way of a reciprocal coincidence of both, just as the mind is both inherent in the body, informing it and connected with corporeal form, and sustained by God, an irradiation as it were, that flows into the body from the divine countenance; whence it [the mind] derives its nobler nature. As this situation establishes the circle as the underlying principle of the harmonious proportions and the source

adumbrat, planum creans; plano vero sectum sphaericum circulum sectione repraesentat, mentis creatae, quae corpori regendo sit praefecta, genuinam imaginem, quae in ea proportione sit ad sphaericum, ut est mens humana ad divinam, linea scilicet ad superficiem, utraque tamen circularis, ad planum vero, in quo et inest, se habet ut curvum ad rectum, quae sunt incommunicabilia et incommensurabilia, inestque pulchre circulus tam in plano secante, circumscribens illud, quam in sphaerico secto, mutuo utriusque concursu, sicut animus et in corpore inest, informans illud connexusque formae corporeae, et in Deo sustentatur, veluti quaedam ex vultu divino in corpus derivata irradiatio, trahens inde nobliorem naturam. Quae causa sicut stabilit proportionibus harmonicis circulum pro subiecto et terminorum

of their determinants, so does it demand the highest possible degree of abstraction because the image of the Mind of God dwells neither in a circle of any given size nor in an imperfect one such as are material and perceptible circles; and, what is the chief thing, because the circle must be kept as free (abstracted) from all that which is material and perceptible as the formulae of the curved line, the symbol of the mind, are separated and, as it were, abstracted from the straight, the simulacrum of the bodies. Thus we are sufficiently prepared for our task of deriving the determinants of the harmonious proportions, subject to the mind alone, from abstract quantities.

fonte, sic vel maxime abstractionem commendat, cum neque in certae quantitatis circulo, neque in imperfecto, ut sunt materiales et sensiles, insit divinitatis animi adumbratio, et quod caput est, tantum a corporeis et sensilibus deceat esse abstractum circulum, quantum curvi rationes, animi symbolum, a recto, corporum umbra, secretae et velut abstractae sunt. Satis igitur muniti sumus ad hoc, ut harmonicis proportionibus, animi solus obiectis, terminos ex abstractis potissimum quantitatibus petamus.[4]

This picture of the relationship between the human mind and the Mind Divine fits in very well with the interpretation of knowledge, already touched upon, as a "matching" of external impressions with pre-existent inner images. Kepler formulates this idea very clearly, adducing his favorite author, Proclus, in support of his views.

For, to know is to compare that which is externally perceived with inner ideas and to judge that it agrees with them, a process which Proclus expressed very beautifully by the word "awakening," as from

Nam agnoscere est, externum sensile cum ideis internis conferre eisque congruum judicare. Quod pulchre exprimit Proclus vocabulo suscitandi, velut e sommo. Sicut enim sensilia foris occurentia faciunt nos recor-

[4] *Harmonices mundi,* Book V (Frisch, Vol. V, p. 223).

sleep. For, as the perceptible things which appear in the outside world make us remember what we knew before, so do sensory experiences, when consciously realized, call forth intellectual notions that were already present inwardly; so that that which formerly was hidden in the soul, as under the veil of potentiality, now shines therein in actuality. How, then, did they [the intellectual notions] find ingress? I answer: All ideas or formal concepts of the harmonies, as I have just discussed them, lie in those beings that possess the faculty of rational cognition, and they are not at all received within by discursive reasoning; rather they are derived from a natural instinct and are inborn in those beings as the number (an intellectual thing) of petals in a flower or the number of seed cells in a fruit is innate in the forms of the plants.

dari eorum, quae antea cognoveramus, sic mathemata sensilia, si agnoscuntur, eliciunt igitur intellectualia ante intus praesentia, ut nunc actu reluceant in anima, quae prius veluti sub velo potentiae latebant in ea. Quomodo igitur irruperunt intro? Respondeo, omnino ideas seu formales rationes harmonicarum, ut de iis supra disserebamus, inesse iis, quae hac agnoscendi facultate pollent, sed non demum introrsum recipi per discursum, quin potius ex instinctu naturali dependere iisque connasci, ut formis plantarum connascitur numerus (res intellectualis) foliorum in flore et cellularum in pomo.[5]

We shall return to Kepler's special views on the morphology of plants. The concept of *instinctus,* which occurs here, is always used by him in the sense of a faculty of perception, whereby he thinks of geometrical forms quantitatively determined. To him, geometry is in fact a value of the highest rank. "The traces of geometry are expressed in the world so that geometry is, so to speak, a

[5] *Harmonices mundi,* Book IV (Frisch, V, p. 224).

kind of archetype of the world."[6] "The geometrical—that is to say, quantitative—figures are rational entities. Reason is eternal. Therefore the geometrical figures are eternal; and in the Mind of God it has been true from all eternity that, for example, the square of the side of a square equals half the square of the diagonal. Therefore, the quantities are the archetype of the world."[7] ". . . the Mind of God, whose copy is here [on earth] the human mind that from its archetype retains the imprint of the geometrical data from the very beginnings of mankind."[8]

I now quote two passages from Book IV, already cited, of the *Harmonices mundi:*

The Christians know that the mathematical principles according to which the corporeal world was to be created are coeternal with God; that God is soul and mind in the most supernally true sense of the word; and that human souls are images of God the Creator, conforming to Him in essentials as well.

. . . rationes creandorum corporum mathematicas Deo coaeternas fuisse Deumque animam et mentem esse superexcellenter, animas vero humanas esse Dei creatoris imagines, etiam in essentialibus suo modo, id sciunt christiani.[9]

[6] *De stella nova,* Ch. IX (Frisch, II, pp. 642 *f.*). ". . . geometriae vestigia in mundo expressa, sic ut geometria sit quidam quasi mundi archetypus."

[7] Letter of Kepler to Hegulontius (Frisch, I, p. 372). "Nobis constat, creatum mundum et quantum factum; geometricae figurae (h.e. quantitativae) sunt entia rationis. Ratio aeterna. Ergo figurae geometricae sunt aeternae, nempe ab aeterno verum erat in mente Dei, lateris tetragonici quadratum, e. gr., esse dimidium de quadrato diametri. Ergo quanta sunt mundi archetypus."

[8] *Apologia contra Fludd* (Frisch, V, p. 429). ". . . in mente divina, cujus exemplar hic est humana, characterem rerum geometricarum inde ab ortu hominis ex archetypo suo retinens."

[9] Book IV, 1, commenting on Proclus (Frisch, V, p. 219).

Mathematical reasoning is "inborn in the human soul" (*eius inerant animae*).

You may ask: how can there be an inborn knowledge of a thing that the mind neither has learned nor ever could learn if it were deprived of the sensory perception of external things? Proclus answered this question in the language constantly used in his philosophy. Nowadays we very properly use, if I am not mistaken, the word "instinct." For quantity is known to the human mind, and to the other souls, by instinct even though it were lacking all the senses for this purpose. The mind is of itself cognizant of the straight line and of an equal interval from one point and can thereby imagine a circle. If the mind can do that, it is even more possible for it to discover proof therein [viz., in the *instinctus*] and thus fulfil the function of the eye in looking at a diagram (if that were necessary). In fact, the mind itself, if it had never possessed an eye, would *demand* an eye in order to comprehend things outside itself and would prescribe the laws of its formation, having obtained them from itself. . . . The very cognition of the quantities, innate in the mind, dictates what the eye ought to be like, and therefore the eye has

Quaeras, qui possit inesse scientia rei, quam nunquam mens didicit nec fortasse discere potest, si sensu rerum externarum destituatur? Ad hoc respondit supra Proclus, verbis in sua philosophia tritis; nos hodie, ni fallor, vocabulo instinctus rectissime utemur. Menti quippe humanae ceterisque animis ex instinctu nota est quantitas, etiamsi ad hoc omni sensu destituatur; illa se ipsa lineam rectam, ipsa intervallum aequale ab uno puncto intelligit, ipsa per haec sibi circulum imaginatur. Si hoc, potest multo magis in eo demonstrationem invenire itaque oculi officium in adspiciendo diagrammate (si tamen opus eo habet) supplere. Quippe mens ipsa, si nullius unquam oculi compos fuisset, posceret sibi ad comprensionem rerum extra se positarum oculum legesque ejus formandi ex se ipsa petitas praescriberet (siquidem pura et sana et sine impedimentis, hoc est si id solum esset, quod est), ipsa enim quantitatum agnitio, congenita menti, qualis oculus esse debeat dictat, et ideo talis

become what it is because the mind is what it is, and not vice versa. But why make many words? Geometry is coeternal with the Mind of God before the creation of things; it is *God Himself* (what is in God that is not God Himself?) and has supplied God with the models for the creation of the world. With the image of God it has passed into man, and was certainly not received within through the eyes.

est factus oculus quia talis mens est, non vicissim. Et quid multis? Geometria ante ortum menti divinae coaeterna, *Deus ipse* (quid enim in Deo, quod non sit ipse Deus?) exempla Deo creandi mundi suppeditavit et cum imagine Dei transivit in hominem, non demum per oculos introrsum est recepta.[10]

When Kepler says, however, that in the Mind of God it has been eternally true that, for example, the square of the side of a square equals half the square of its diagonal, we do not, to be sure, begrudge one of the first joyful discoverers of quantitative, mathematically formulated natural laws his elation but must, as modern men, remark in criticism that the axioms of Euclidian geometry are not the only possible ones. I have already set up a signboard warning that theses determined by means of rational formulations should never be declared the only possible premises of human reasoning. In that connection I had particularly in mind certain formulations of Kant's philosophy that seem to me to be misleading. I therefore propose to leave, even with respect to geometry, the *a priori* at the metaphorical preliminary stage of ideas guiding the *instinctus* (which is the reason why I cannot follow the scholar who has translated Kepler's word *instinctus* by "reine Anschauung"). On the other hand, I entirely

[10] "De configurationibus harmonis" (pp. 222 *ff.*).

share the opinion that man has an instinctive tendency, not rooted merely in external experience, to interpret his sensory perceptions in terms of Euclidean geometry. It took a special intellectual effort to recognize the fact that the assumptions of Euclidean geometry are not the only possible ones. Probably even the modern thinker can agree with the following general formulation of Kepler's: "The perceptible harmonies have that in common with the archetypal harmonies that they require terms and the comparison of terms, the activity (*energeia*) of the soul itself; in this comparison consists the essence of both." [11]

4

We now follow Kepler one step further down in the hierarchical order of his universe, passing, that is, from the ideas in the mind of the Godhead to the corporeal world. Here the heavenly bodies, with the sun as the central point in the sense of the *signatura rerum,* are for him a realization of the ideal, spherical image of the Trinity though less perfect than it. The sun in the centre, as the source of light and warmth and accordingly of life, seems to him especially suited to represent God the Father. I quote on this point the following very typical passage from his book on optics:

[11] *Harmonices mundi,* Book IV (Frisch, V, p. 223). "Commune enim habent harmoniae sensiles cum archetypalibus, quod terminos requirant eorumque comparationem, ipsius animae energiam; in hac comparatione utrarumque essentia consistit."

First of all the nature of every thing was bound to represent God its creator as far as it was able to do so within the condition of its being. For when the all-wise Creator sought to make everything as good, beautiful, and excellent as possible, He found nothing that could be better or more beautiful or more excellent than Himself. Therefore when He conceived in His Mind the corporeal world He chose for it a form that was as similar as possible to Himself. Thus originated the entire category of the quantities, and within it the differences of the curved and the straight, and the most excellent figure of all, the spherical surface. For in forming this the most wise Creator created playfully the image of His venerable Trinity. Hence the centre point is, as it were, the origin of the spherical body; the outer surface the image of the innermost point, as well as the way to arrive at it; and the outer surface can be understood as coming about by an infinite expansion of the point beyond itself until a certain equality of all the individual acts of expansion is reached. The point spreads itself out over this extension so that point and surface are identical, except for the fact that the ratio of den-

Primum omnium rerum natura Deum conditorem, quantum quaeque suae essentiae conditione potuit, repraesentare debuit. Nam cum Conditor sapientissimus omnia studeret quam optima, ornatissima, praestantissimaque efficere: nihil seipso melius ornatiusque, nihil praestantius repperit. Propterea cum corporeum mundum agitaret animo, formam ei destinavit sibi ipsi quam simillimam. Hinc ortum totum quantitatum genus, et in eo curvi rectique discrimina, praestantissimaque omnium figura, Sphaerica superficies. Nam in ea formanda lusit sapientissimus Conditor adorandae suae Trinitatis imaginem. Hinc Centri punctum, est Sphaerici quaedam quasi origo, superficies puncti intimi imago, et via ad id inveniendum, quaeque infinito puncti egressu ex se ipso, usque ad quandam omnium egressuum aequalitatem, gigni intelligitur, puncto se in hanc amplitudinem communicante, sic ut punctum et superficies, densitatis cum amplitudine commutata proportione, sint aequalia: Hinc est undique punctum inter et superficiem absolutissima aequalitas, arctissima unio, pulcherrima conspiratio, connexus, relatio, proportio, commensus. Cumque Tria sint plane Centrum, Superficies et Interval-

sity and extension is reversed. Hence there exists everywhere between point and surface the most absolute equality, the closest unity, the most beautiful harmony [literally: breathing together!], connection, relation, proportion, and commensurability. And although Centre, Surface, and Distance are manifestly Three, yet are they One, so that no one of them could be even imagined to be absent without destroying the whole.

lum; ita tamen unum sunt, ut nullum ne cogitatu quidem abesse possit, quin totum destruatur.

This, then, is the genuine and most suitable image of the corporeal world, and every being among those physical creatures that aspire to the highest perfection assumes it [viz., the spherical shape] either absolutely or in a certain respect. Therefore the bodies themselves which, as such, are confined by the limits of their surfaces and are thus unable to expand into spherical form, are endowed with various powers, nesting in the bodies, which are somewhat freer than the bodies themselves, possessing no corporeal matter but consisting of a particular matter of their own that assumes geometrical dimensions, and which powers flow out from them and aspire to the circular form—as can be clearly seen, especially, in the magnet but also in many other things. Is it any wonder, then,

Haec igitur genuina, haec aptissima corporei mundi est imago quam vel simpliciter vel respectu quodam suscipit, quicquid ad summam perfectionem inter corporeas creaturas aspirat. Propterea corpora ipsa, cum per sese suarum superficierum finibus continerentur, nec se ipsa multiplicare possent in orbem, variis sunt praedita virtutibus, quae nidulantes quidem in corporibus, seipsis vero paulo liberiores, et materia carentes corporea, sed sua quadam constantes materia, quae dimensiones suscipit Geometricas, egrederentur, orbemque adfectarent: ut praecipue in Magnete, sed et in multis aliis clare apparet. Quid mirum igitur, si

if that principle of all beauty in the world,[12] which the divine Moses introduces into scarcely created matter, even on the first day of creation, as (so to speak) the Creator's instrument, by which to give visible shape and life to all things—is it any wonder, I say, if this primary principle and this most beautiful being in the whole corporeal world, the matrix of all animal faculties, and the bond between the physical and the intellectual world, submitted to those very laws by which the world was to be formed? Hence the sun is a certain body in which [resides] that faculty of communicating itself to all things which we call light. For this reason alone its rightful place is the middle point and centre of the whole world, so that it may diffuse itself perpetually and uniformly throughout the universe. All other beings that share in light imitate the sun.

principium illud omnis in mundo ornatus, quod divinus Moses quasi quoddam Creatoris instrumentum, ad figuranda et vegetanda omnia, die statim primo in materiam vix conditam introducit: si hoc inquam principium, et res in toto corporeo mundo praestantissima, matrix animalium facultatum, vinculumque corporei et spiritualis mundi, in leges easdem transiverit, quibus mundus erat exornandus. Sol itaque corpus est quodpiam, in eo haec sese rebus omnibus communicandi facultas, quam lucem appellamus; cui vel ob hanc causam medius in toto mundo locus, et centrum debetur, ut aequabiliter perpetuo sese in Orbem totum diffunderet. Solem omnia alia, quae lucis sunt participia, imitantur.[13]

I should, first of all, like to point out that Kepler alludes here to a photometric law expounded by him in this book, according to which, as we say today, the intensity of illumination decreases in inverse ratio to the square of the distance from the source of light, which is conceived as a point. The word *amplitudo,* translated by

[12] The sun.

[13] *Ad Vitellionem paralipomena,* pp. 6–7.

"extension," obviously means here the area of the sphere's surface, which is, of course, proportional to the square of the length of the radius. This photometric law of Kepler's is of great importance and brought him very close to the discovery of the law of gravitation. From this example it can be seen that in Kepler the symbolical picture precedes the conscious formulation of a natural law. The symbolical images and archetypal conceptions are what cause him to seek natural laws. For this reason we also regard Kepler's view of the correspondence between the sun with its surrounding planets and his abstract spherical picture of the Trinity as primary: *because he looks at the sun and the planets with this archetypal image in the background he believes with religious fervour in the heliocentric system*—by no means the other way around, as a rationalistic view might cause one erroneously to assume. This heliocentric belief, to which Kepler remained faithful from early youth, impels him to search for the true laws of the proportion of planetary motion as the true expression of the beauty of creation. At first this search went in a wrong direction, which was later corrected by the results of actual measurement.

Kepler's conception of the sun with the planets as the image of the Trinity is also very clearly revealed in the following quotation from his treatise, *Tertius interveniens,* written in German. We shall dwell later on the significance of the title and the other contents of the book. The passage in question is taken from Section 126, which bears the title "A Philosophical Discourse *de signaturis rerum.*" It runs thus:

And as the heavenly *corpora* (*orbes*) are *vel quasi* signified and depicted in the geometrical *corporibus* and *contra:* So also will the heavenly movements that take place in a *circulo* correspond to the geometrical *planis circulo inscriptis.* (See above, num. 59.)

Indeed, the most holy Trinity is depicted in a *sphaerico concavo,* and this in the world, and *prima persona, fons Deitatis, in centro;* the *centrum,* however, is depicted in the sun, *qui est in centro mundi;* for it, too, is a source of all light, movement, and life in the world.

Thus is *anima movens* represented *in circulo potentiali* which is *in puncto distincto:* thus a physical thing, a *materia corporea,* is represented in *tertia quantitatis specie trium dimensionum:* thus is *cuiusque materia forma* represented *in superficie.* For as a *materia* is informed by its *forma,* so is a geometrical *corpus* shaped by its external facets and *superficies:* of which things many more could be adduced.

Now, as the Creator played, so he also taught Nature, as His image, to play; and to play the very same game that He played for her first. . . .

From these words of simple beauty it appears that Kepler connects the Trinity with the three-dimensionality of space and that the sun with the planets is regarded as a less perfect image of the abstract spherical symbol. By means of this conception, which is related to the idea of correspondences, Kepler avoids a pagan worship of Helios and remains true to Christian belief. In this connection I should also like to mention the "Epilogus de Sole conjecturalis" with which Kepler concludes his chief work, the *Harmonices mundi,* and in which, among other things, he defines, from his Christian point of view, his attitude toward the pagan hymn to the sun by his favorite author, Proclus. Kepler's notion of a playful activity established ever since the creation of the world and replayed by nature in imitation of the original is also in accord with the idea of the "signature."

With regard to the concept "anima movens" I should like to remark that Kepler's views on the cause of movement are vacillating. In one passage in his treatise on the new star he says:

Finally, those motive powers of the stars share in some way in the capacity of thought so that as it were they understand, imagine, and aim at their path, not of course by means of ratiocination like us human beings but by an innate impulse implanted in them from the beginning of Creation; just so do the animal faculties of natural things acquire, though without ratiocination, some knowledge of their goal to which they direct all their actions.

Denique ut facultates illae stellarum motrices sunt mentis quodammodo participes, ut suum iter quasi intelligant, imaginentur, affectent, non ratiocinando quidem, ut nos homines, sed ingenita vi et quae in prima creatione ipsis est instincta: sic facultates animales rerum naturalium obtinent quendam intellectum finis sui (sine quidem ratiocinatione) in quem omnes suas actiones dirigunt.[14]

Here Kepler adopts the animistic point of view. But elsewhere he says:

The sun in the midst of the movable stars, itself at rest and yet the source of motion, bears the image of God the Father, the Creator. For what creation is to God motion is to the sun; it moves, however, within the fixed stars as the Father creates in the Son. For if the fixed stars did not create space by means of their state of immobility, nothing could move. The sun, however, distributes its motive force through the me-

Sol in medio mobilium quietus ipse et tamen fons motus, gerit imaginem Dei patris creatoris. Nam quod est Deo creatio, hoc est Soli motus, movet autem in fixis, ut Pater in Filio creat. Fixae enim nisi locum praeberent sua quiete, nihil moveri posset. Dispertitur autem Sol

[14] *De stella nova,* Ch. XXVIII (Frisch, II, p. 719).

dium in which the movable things exist, just as the Father creates through the Holy Ghost or through the power of the Holy Ghost.	virtutem motus per medium, in quo sunt mobilia, sicut Pater per Spiritum vel virtute Spiritus sui creat.[15]

This conception has much in common with modern physics of fields. As a matter of fact Kepler thought of the gravitation emanating from the sun as similar to light and yet differing from it. He also compares this gravitation to the effect of magnets, with reference to Gilbert's experiments.

In view of Kepler's conflict with Fludd, the representative of traditional alchemy, a conflict which we shall discuss later, it is important that Kepler's symbol—of a type designated by C. G. Jung as a mandala because of its spherical form—contains no hint of the number four or quaternity. This is all the more significant since Kepler had an excellent knowledge of the Pythagorean numerical speculations, particularly of the *tetraktys,* which he discusses in detail in the third book of his *Harmonices mundi* as an historical introduction to his own theory of musical intervals. But these ancient speculations are to Kepler a mere curiosity; for him the number four has no symbolical character. Perhaps the lack of a symbolism of time in Kepler's spherical picture is related to the lack of any suggestion of quaternity. Movement in a straight line, directed away from the centre, is the only kind contained in Kepler's symbol, and in so far as this movement is caught up by the outer surface of the sphere the symbol can be termed static. Since the Trinity had never been represented in this particular way before Kepler, and

[15] *Epistola ad Maestlinum* (Frisch, I, p. 11).

since he stands at the threshold of the scientific age, one is tempted to assume that Kepler's "mandala" symbolizes a way of thinking or a psychological attitude which, far transcending Kepler's person in significance, produced that natural science which we today call classical. From within an inner centre the psyche seems to move outward, in the sense of an extraversion, into the physical world in which, by definition, everything that occurs is automatic; so that the mind, itself in a state of rest, embraces this physical world, as it were, with its ideas.[16]

5

The next step in Kepler's hierarchical arrangement of the cosmos, which we have already traced from the trinitarian Godhead and the ideas in the Mind of God through their

[16] According to the psychology of Jung the psychological processes that accompany an enlargement of consciousness can be represented as the coming into being of a new *centre* embracing conscious as well as unconscious contents (called "self" by Jung). These "centring" processes are always characterized by the symbolical pictures of the mandala and of rotating movement. In Chinese texts the latter is very vividly termed "circulation of light."

In an attempt to apply these results of analytical psychology to the phase of intellectual history known as the rise of classical mechanics in the seventeenth century (which is most closely connected with the heliocentric idea), we should bear in mind that the attention of the scientists who helped to found classical mechanics was directed only outward. It is therefore to be expected that the above mentioned inner centring processes, together with the appropriate images, would be projected outward. Indeed we can observe, in Kepler's views specifically, that the planetary system with the sun as centre became the bearer of the mandala-

spherical model down to the physical world, the sun, and the heavenly bodies revolving about it, lead us now to the *individual souls.*

We have already said that for Kepler the earth is a living thing like man. As living bodies have hair, so does the earth have grass and trees, the cicadas being its dandruff; as living creatures secrete urine in a bladder, so do the mountains make springs; sulphur and volcanic products correspond to excrement, metals and rainwater to blood and sweat; the sea water is the earth's nourishment. As a living being the earth has a soul, the *anima terrae,* with qualities that can be regarded as to a large extent analogous to those of the human soul, the *anima hominis.*[17] We can therefore understand by individual soul the *anima terrae* and the souls of the planets as well as the

picture, the earth being related to the sun as is the ego to the more embracing "self." It appears that in this way the heliocentric theory received, in the minds of its adherents, an injection of strongly emotional content stemming from the unconscious. Perhaps the projection of the above mentioned symbolical image of the inner rotating movement onto the external rotation of the heavenly bodies contributed to the investment of this external rotation with an absolute validity that went beyond empirical experience. An additional argument for this opinion is that in Newton the ideas of absolute space and absolute time entered even into his theological views.

[17] The conception of the earth as a living being with a soul is already present in late antiquity. See on this point: Cicero, *De natura deorum,* II, 83; Ovid, *Metamorphoses,* XV, 342; Seneca, *Quaestiones naturales,* VI, 16, 1; Plotinus, IV, 4. See also the article "Plotinus" by H. R. Schwyzer in A. Pauly, G. Wissowa, W. Kroll, *Real-Encyklopädie der klassischen Altertumswissenschaft* (1951 edn.), XI, cols. 471–592. Note especially col. 578, where the idea of the animation of the earth is traced back to Posidonius.

human souls. At the same time the *anima terrae* is also a formative power (*facultas formatrix*) in the earth's interior and expresses, for example, the five regular bodies in precious stones and fossils. In this Kepler follows views also held by Paracelsus. The latter had employed the concept of the "Archaeus" as a formative principle of nature which, as *signator,* also creates the *signaturae.* As a matter of fact Kepler admitted in his dispute with Fludd that the latter could also call the *anima terrae* "Archaeus" if he preferred that.[18] It is important that in Kepler's view the *anima terrae* is responsible for the weather and also for meteoric phenomena. Too much rain, for instance, is an illness of the earth.

Now Kepler's characteristic basic idea concerning the individual soul is that it is, as an image of God however imperfect, partly a point and partly a circle: "anima est punctum qualitativum." This theory goes back to neo-Platonic and neo-Pythagorean philosophers of late antiquity, in whose works similar ideas can be found.[19]

[18] Frisch, V, p. 440.

[19] According to Sextus Empiricus (*Adversus mathematicos,* III, 22), the point (στιγμή) is γεννητική and not incorporeal; it is the monad and the soul.

According to Plotinus, IV, 4, 16, the soul is fitted like a circle to its own centre, hence closely joined to the centre, an unextended extension: καὶ ἡ ψυχὴ ἡ τοιαύτη, οἷον κύκλος προσαρμόττων κέντρῳ εὐθὺς μετὰ κέντρον αὐξηθείς, διάστημα ἀδιάστατον.

On the concept *Inlocalitas animae* (ἀδιάστατος or ἀδιαστασία) in Claudianus Mamertus cf. also the essay by E. Bickel, "*Inlocalitas,* Zur neupythagoreischen Metaphysik," in *Immanuel Kant* (studies in honour of Kant's 100th birthday; Leipzig, 1924); further, F. Bömer, *Der lateinische Neuplatonismus und Neu-*

Which functions of the soul are attributed to the central point and which to the peripheral circle is a somewhat doubtful matter. In general the contemplative and imaginative functions are assigned to the point, the active and motor effects on the body to the circle. The latter, however, is also supposed to correspond to the faculty of *ratiocinatio,* reflection and logical conclusion. The process of the issuing forth of the soul from the central point to the periphery of the circle is often compared by Kepler to the emanating of a flame. He also emphasizes expressly the analogy of this movement to the rays of light streaming from the sun as centre, thus also establishing a connection with the radii that issue from the central point in his symbol of the Trinity. It is not difficult to see this process of the emanation of the soul from the point to the circle as analogous to the extraversion of modern psychology, from the point of view of which creation, in Kepler's system of ideas, would be the divine model, whereas the being of God would have to be regarded as the model of introversion.

The following passage from *Harmonices mundi* should make Kepler's view clear:

Firstly, the soul has the structure of a point in actuality (at least by reason of its conjunction with its body), and the figure of a circle in potentiality. Now, since it is energy, it pours itself forth from that punctiform abode into a circle.

Primum anima puncti rationem sortita est actu (saltem ratione alligationis ad corpus suum), circuli figuram potestate; quae cum sit energia, edidit sese ab illa sede puncti in circulum; sive enim sentire debeat res externas, illae sphaericum in mo-

pythagoreismus und Claudianus Mamertus in Sprache und Philosophie (Leipzig, 1936), especially pp. 124 and 139.

Whether it is obliged to perceive external things that surround it in spherical fashion or whether it must govern the body (the body, too, lies round about it), the soul itself is hidden within, rooted in its fixed point whence it goes out into the rest of the body by a semblance of itself. But how should it go out if not in a straight line (for that is truly a "going out")? How should it have any other way of going out, being itself both light and flame, than as the other lights go out from their sources, that is, in straight lines? It goes out, then, to the exterior of the body according to the same laws by which the surrounding lights of the firmament come in towards the soul that resides in a point.

dum sese circumstant, sive corpus regere, corpus quoque circumjectum habet, ipsa latet intus, radicata in puncto eius certo, unde exit per speciem sui in corpus reliquum. At qui exeat, nisi per lineas rectas? hoc enim vere est exire; qui alium exeundi morem habeat, ipsa et lux existens et flamma, quam a fontibus suis exeunt alia lumina, lineis sc. rectis? Egreditur igitur versus exteriora corporis iisdem legibus, quibus circumstantia firmamenti lumina versus illam in puncto residentem ingrediuntur.[20]

With this conception of the soul both as point and circle are connected Kepler's special views on astrology, to the discussion of which he specifically devoted the treatise *Tertius interveniens*. Here Kepler intervenes as a third party in the dispute between H. Röslinus, representing the point of view of traditional astrology, and P. Feselius, who disparages all astrology as superstition, in order to oppose to both authors his own essentially divergent point of view. On the first leaf of the book, immediately after the title, appears the commentary: "A warning to sundry Theologos, Medicos, and Philosophos, in particular to D. Philippus Feselius, that they should not,

[20] Book IV (Frisch, V, p. 258).

in their just repudiation of star-gazing superstition, throw out the child with the bath and thus unknowingly act in contradiction to their profession." Kepler also formulates his ideas on astrology in his earlier treatise on the new star, arguing there against Pico della Mirandola, and, in conclusive fashion, reverts once more to the subject in his chief work, *Harmonices mundi.*

In what follows we shall first attempt—setting aside the question of the objective validity of astrological statements—to characterize Kepler's integration of his own astrology, so different from the usual kind, into the whole system of his ideas on natural science, which are what interests us here.

According to Kepler, the individual soul, which he calls *vis formatrix* or *matrix formativa,* possesses the fundamental ability to react with the help of the *instinctus* to certain harmonious proportions which correspond to specific rational divisions of the circle. In music this intellectual power reveals itself in the perception of euphony (consonance) in certain musical intervals, an effect that Kepler thus does not explain in a purely mechanical way. Now the soul is said to have a similar specific reactability to the harmonious proportions of the angles which the *rays* of stellar light, striking the earth, form with each other. It is with these, in Kepler's opinion, that astrology should concern itself. According to him, then, the stars exert no special remote influence, since their true distances are of no importance to astrology and only their light rays can be regarded as effective. The soul knows about the harmonious proportions through the *instinctus* without conscious reflection (*sine ratiocinatione*) because the soul, by virtue of its circular form, is an image of God in Whom these proportions and the geometric truths following therefrom exist from all eternity. Now since the soul, in consequence of its circular form, has knowledge of these, it is impressed by the external forms of the configurations of rays and retains a memory of them from its very birth. I cite Kepler's words on this:

I speak here as do the astrologers. If I should express my *own* opinion it would be that there is no evil star in the heavens, and this, among others, chiefly for the following reasons: it is the nature of man as such, dwelling as it does here on earth, that lends to the planetary radiations their effect on itself; just as the sense of hearing, endowed with the faculty of discerning chords, lends to music such power that it incites him who hears it to dance. I have said much on this point in my reply to the objections of Doctor Röslin to the book *On the New Star* and in other places, *passim,* also in Book IV of the *Harmonices, passim,* especially Chapter 7.

Loquor cum astrologis. Nam si *meam* sententiam dicam, nullus in coelo maleficus mihi censetur: idque cum ob alias rationes tum maxime propter hanc quia hominis ipsius natura est hic in Terra versans, quae radiationibus planetarum conciliat effectum in sese; sicut auditus, instructus facultate dignoscendi concordantias vocum conciliat musicae hanc vim, ut illa incitet audientem ad saltandum. De hac re egi multis in responso ad objecta Doctoris Roeslini contra libellum de stella nova et alibi passim, etiamque in lib. IV Harmonicorum passim, praesertim cap. 7.[21]

For, the *punctum naturale* (the natural soul in every human being or also in the terrestrial globe itself) has as much power as a real *circulum. In puncto inest circulus in potentia propter plagas unde adveniunt radii se mutuo in hoc puncto secantes.*[22]

[21] *Mysterium cosmographicum* (Frisch, I, p. 133): "In Caput nonum notae autoris" (in the 2nd edn. of the book).

[22] *Tertius interveniens,* No. 40.

The natural soul of man is not larger than a single point; and upon these points the shape and character of the whole heaven, be it a hundred times as large as it is, are imprinted *potentialiter*.[23]

The nature of the soul behaves like a point; for this reason it can also be transformed into the points of the *confluxus radiorum*.[24]

The soul, according to Kepler, contains the idea of the zodiac within itself by virtue of its inherent circular form; but it is the planets, and not the fixed stars, which (through the intermediary of light) are the effective vehicles of astrological influence. The "distribution of the twelve signs among the seven planets" is for him a fable; yet the *doctrina directionum,* he thinks, is based on good reason since it emphasizes the harmonious combination of two rays of light that is called an "aspect."

In *Harmonices mundi,* Kepler expresses this with particular clarity:

Inasmuch as the soul bears within itself the idea of the zodiac, or rather of its centre, it also feels which planet stands at which time under which degree of the zodiac, and measures the angles of the rays that meet on the earth; but inasmuch as it receives from the irradiation of the Divine essence the geometrical figures of the circle and (by comparing the circle with certain parts of it) the archetypal harmonies (not, to be sure, in purely geometri-

Quatenus igitur haec anima circuli zodiaci seu potius eius centri gestat ideam, persentiscit etiam, qui planeta quovis tempore sub quo zodiaci gradu versetur angulosque radiationum, coeuntium in Terra, metitur; quatenus vero ex Divinae essentiae irradiatione rationes circuli geometricas et (per circuli comparationem cum certis suis partibus) harmonias archetypales suscepit, non pure quidem geometricas, sed radiationum lucidarum veluti saccaro quodam

[23] No. 42.

[24] No. 65.

cal form but as it were overlaid or rather completely saturated with a filtrate of glittering radiations), it also recognizes the measurements of the angles and judges some as congruent or harmonious, others as incongruent.	inductas, imo penitus imbutas, mensuras etiam angulorum jam agnitas, has congruas seu harmonicas, illas incongruas iudicat.[25]

The human soul, in Kepler's view, flows at birth into a pre-existent form, which is shaped on the earth by these rays of light from the stars (planets).

Cf. the *Tertius interveniens:*

For it is by no means to be pronounced foolishness that man is *naturali necessitate* diversified and qualified in accordance with the *Configurationibus stellarum;* this might really far rather be called an "influence" of the nature of man into the star (as of fluid plaster into a mould) than, on the contrary, an "influence" of the star into man.[26]

The effective angles between two rays of light are, according to Kepler, those that correspond either to the regular polygons with which a plane can be covered without gaps, such as equilateral triangles, squares, or hexagons; or to star-shaped figures that bear a close relation to the regular polyhedra. Here Kepler tries to establish an intimate connection with the proportions appearing in the musical intervals of consonances; but he is forced to admit also certain differences between these and the astrologically effective divisions of the circle. I shall not go into the details of this but shall merely reproduce some figures from the *Harmonices mundi.* In them (Fig. 1) can be seen the reciprocal connection of the circumferential

[25] Book IV (Frisch, V, p. 256).
[26] No. 107.

figure with a central figure, this connection being such that the peripheral angle between two adjacent sides of the latter is equal to the central angle between the radii to adjacent points of the former, and vice versa.

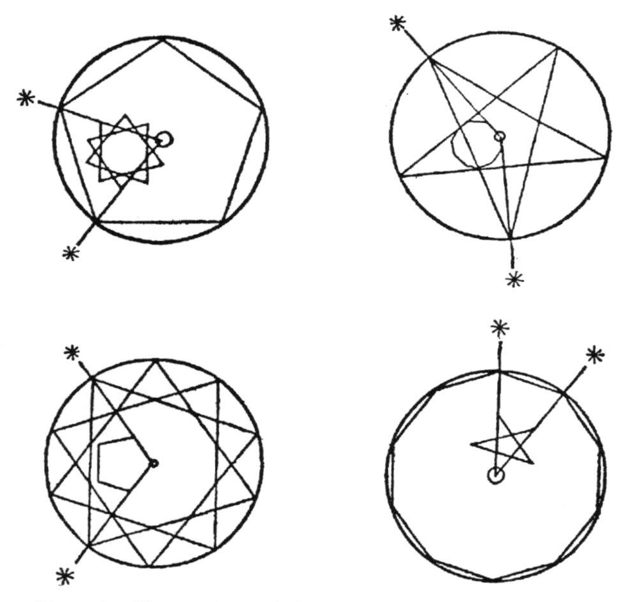

FIG. 1. Circumferential figure and central figure

From Kepler's *Harmonices mundi,* Book IV: *De configurationibus harmonicis,* Ch. 5 (Frisch, V, pp. 238 and 239: figs. 32, 33, 34, 35).

In Kepler's view, the two figures are supposed to correspond to the circular form and the point form of the soul. I give the following passage at some length, since these ideas may perhaps be of particular psychological interest.

The same thing can be proved from the inner properties of the soul, already touched upon in Chapter 3. For, since it is from the soul that the harmonies of the configurations obtain their formal being (*esse formale*), the soul certainly possesses an intimate knowledge of the figures, both circumferential and central; [this knowledge it possesses] by virtue of the same distinction by which the soul is both a circle and a point, that is, the centre of a circle. But although every soul bears within itself a certain idea of the circle—a circle not merely detached from matter but also somehow from magnitude (as has been said in Chapter 3), so that in this case circle and centre almost coincide and the soul itself can be called a potential circle as well as a point differentiated according to directions and thus somehow qualified—nevertheless there must be observed the distinction that some faculties of the soul have to be considered rather as circle, others as point. For as one cannot imagine a circle without a centre and, conversely, the point has about it an area circumscribable by a circular line, so there is no activity in the soul without an impression upon the imagination. Conversely, all inter-

Idem etiam sic probatur ex intimis animae proprietatibus, cap. 3 tactis. Cum enim anima sit, quae configurationum harmoniis suum conciliat esse formale, certe quo discrimine anima vel circulus est vel punctum, centrum circuli, eodem discrimine etiam familiares illi erunt figurae, circumferentialis et centralis. Etsi vero omnis anima circuli quandam ideam gerit, abstracti quidem illius non tantum a materia, sed etiam a magnitudine quodammodo, ut dictum cap. 3, eoque circulus et centrum hic fere coincidunt ipsaque vel circulus potentialis, vel punctum plagis distinctum et sic quodammodo qualitativum dici potest; tamen discrimen hoc videtur observandum, quod aliae facultates potius ut circulus considerandae sunt, aliae potius ut punctum. Quemadmodum enim circulus sine centro cogitari nequit, omne vicissim punctum circa se habet regionem scribendo circulo, sic in anima quoque operatio nulla est sine impressione imaginativa, omnis vicissim in-

nal reception or meditation is caused by external movement, every inward function of the soul by outward movements. The principal and highest faculty of the soul, which is called mind, what is it if not the centre? The ratiocinative faculty, what is it if not the circle? For as the centre is within and the circle without, so the mind [*mens*] remains within itself, whereas ratiocination weaves a sort of outer covering; and as the centre is the basis, source, and origin of the circle, so is the mind of ratiocination.

On the other hand, all these faculties of the soul—the mind, the faculty of ratiocination, and even the sensitive faculty—are a sort of centre whereas the motor functions of the soul are the periphery; for again, as the outer circle is drawn around the centre, so is action directed outward whereas cognition and meditation are performed inwardly; and as the circle is related to the point so is outward action to inner contemplation, and animal movement to sensory perception. The point, namely, because it is everywhere opposite to the circumference, is by its very nature suited to represent the passively receptive; and the sensitive soul—that which is perceptive of the rays of the

terna receptio vel meditatio est propter motum externum, omnis animae facultas interior propter magis [should read: motus] exteriores. Ipsa princeps et suprema animae facultas, mens dicta, quid est nisi centrum? quid ratiocinativa, nisi circulus? Nam sicut centrum intus est, circulus exterius, sic mens secum ipsa manet, ratiocinatio telam quandam exteriorem texit; et sicut centrum circuli, sic mens ratiocinationum basis, fons et origo est.

Rursum omnis haec animae facultas tam intellectus, quam discursus, denique etiam sensitiva, sunt centrum quoddam, at facultates animae motrices, circulus; quia rursum ut circulus externus circumponitur centro, sic operatio ad extra est, cognitio meditatioque perficitur intus, et ut circulus ad punctum, sic quodammodo se habet actio externa ad contemplationem internam, motus animalis ad sensionem. Punctum enim, quia undique oppositum circumferentiae, aptum natum est repraesentando patienti, et anima sensitiva vel haec radiationum perceptiva, quid aliud sentiendo et percipiendo, quam patitur? sc. quia movetur objectis. Comparando etiam utramque comparationem: ut

planets—what else does it do in sensing and perceiving than be passive, that is to say, be moved by that which is opposite to it? Now compare the above two comparisons with each other: as the central point is the same in both cases, so also is the form of cognition somehow the same, the chief, mental [form of cognition] and the sensory one or rather its analogue which perceives the radiations. None of these [cognitive functions] makes, as such, use of discursive ratiocination but has knowledge without it. Thus *this*—I mean, sublunary nature and also the sensitive nature [in man]—is a slight image of *that*, to wit, the principal [faculty], the human mind: just so is *that* logical reasoning an image of *these* actions or operations of the soul, and either is a circle.

idem utrinque centrum, sic etiam eadem quodammodo cognitionis forma est, mentalis princeps et sensitiva vel ei analoga, perceptiva radiationum; neutra discursu in se ipsa, quatenus talis, utitur, sed cognoscit citra illum. Ut ita sit haec illius, natura dico sublunaris aut etiam sensitiva, mentis humanae principis tenuis quaedam imago, sicut ille discursus rationis harum actionum aut operationum animae imago est, utraque circulus.

In so far, then, as the souls are perceptive of the celestial radiations and are thus moved by them, as it were, with an inward, self-contained movement, we must regard them as points; but in so far as they in turn cause movement, that is to say, transfer the harmonies of the radiations which they have perceived into their operations and are stimulated to action by them, they ought to be considered as circles. It follows that,

Quatenus igitur animae percipiunt radiationes coelestes et sic iis quasi moventur secum ipsae intus, nobis puncta sunto, quatenus vero vicissim movent, hoc est quatenus perceptas radiationum harmonias transferunt in opera sua iisque stimulantur ad agendum, considerari debent ut circulus. Se-in so far as the soul takes cognizance of the harmonies of the rays, it must chiefly concern itself with the central figure; but in so far as it acts, provoking meteoric phenomena (or what corresponds to these in human beings) it must devote itself to the circumferential figure. In an aspect, however, the effectiveness [27] is of greater interest to us than the manner in which the aspect may be perceived by the operating soul; therefore the consideration of the circumferential figure is more important to us than that of the central figure.

quitur igitur, ut in quantum cognoscit harmonias radiorum, occupetur potissimum circa centralem figuram; in quantum vero operatur, ciens meteora (et quae similia in homine), circumferentiali sese accomodet. Et vero in aspectu prior est nobis cura efficaciae quam modi, quo is percipiatur ab anima operante, prior igitur et circumferentialis quam centralis figurae respectus.[28]

So much for the inner and the outer figure of the aspects; the greater importance attributed to the outer figure by Kepler seems to indicate once more a predominantly extravert attitude. Since the *anima terrae* causes the weather and, like everything partaking of the nature of the soul, has the faculty of reacting to the aspects, the weather must be sensitive to these aspects. Kepler is convinced that he has proved this in numerous reports on the weather, and he then, conversely, regards this as proof of the existence of the *anima terrae*. The animistic conception of the cause of planetary movement, of which we have already spoken, leads Kepler to the assumption of a universal connection between the phenomena of the heavens and the receptive faculties of the individual souls.

[27] As defined in Book IV, Ch. 5 (Frisch, V, page 235).

[28] *Harmonices mundi,* Book IV, Proposition VI (Frisch, V, p. 238).

Nothing exists or happens in the visible heavens the significance of which is not extended further, by way of some occult principle, to the earth and the faculties of the natural things; and thus these animal faculties are affected here on earth exactly as the heavens themselves are affected.

Nihil esse vel fieri in coelo visibili, cuius sensus non occulta quadam ratione in Terram inque facultates rerum naturalium porrigatur: easque facultates animales sic affici hic in Terris, ut coelum ipsum afficitur.[29]

It is interesting that Kepler tries to supplement the passive, receptive manifestation of the *vis formatrix* by an active effect of the same *vis formatrix* in making it responsible for the morphology of the plants. Whatever is sensitive to harmonious forms can also produce harmonious forms, such as, for example, the blossoms of plants with their regular number of petals, and vice versa. He therefore raises the question as to whether the vegetative soul of plants, too, has the ability to react to the proportions of the planetary rays but leaves it unanswered because he will not make any assertions without having performed experiments of his own.

It is apparent from what has been said above that in Kepler's theoretical ideas astrology has been completely integrated with scientific-causal thinking; in strongly emphasizing the role of the light rays he made it a part of physics and, indeed, of optics. The astrological effectiveness of directions that are geometrically defined in relation to the sphere of the fixed stars but do not coincide with light rays (as, for example, the direction from the earth to the vernal point) is expressly rejected by Kepler. Furthermore, he stresses again and again the fact that in

[29] *De stella nova,* Ch. 28 (Frisch, II, p. 719).

his view astrological effects are not caused by the celestial bodies but rather by the individual souls with their specifically selective reactability to certain proportions. Since this power of reacting, on the one hand, receives influences from the corporeal world and, on the other hand, is based on the image relation to God, these individual souls (the *anima terrae* and the *anima hominis*) become for Kepler essential exponents of cosmic harmony (*harmonia mundi*).

Kepler's peculiar conception of astrology met with no recognition. In fact, if one proceeds on this basis it hardly appears possible to avoid the empirically untenable conclusion that artificial sources of light would also be able to produce astrological effects. In general, I should like to remark in criticism of astrology that, in consequence of the vague character of its pronouncements (including the famous horoscope that Kepler drew up for Wallenstein), I see no reason to concede to horoscopes any objective significance independent of the subjective psychology of the astrologer.[30]

6

Kepler's views on cosmic harmony, essentially based on quantitative, mathematically demonstrable premises, were incompatible with the point of view of an archaic-magical description of nature as represented by the chief work of a respected physician and Rosicrucian, Robert Fludd

[30] On this point, cf. also the negative result of the statistical experiment described by C. G. Jung in Ch. 2 of his contribution to this volume.

of Oxford: *Utriusque Cosmi Maioris scilicet et Minoris Metaphysica, Physica atque technica Historia,* first edition, Oppenheim, 1621. In an appendix to Book V of the *Harmonices mundi* [31] Kepler criticized this work of Fludd's very violently. Fludd, as the representative of traditional alchemy, published in his treatise *Demonstratio quaedam analytica* [32] a detailed polemic directed against Kepler's appendix, whereupon the latter replied with an *Apologia* [33] that was followed by a *Replicatio* [34] from Fludd.

The intellectual "counter-world" with which Kepler here clashed is an archaistic-magical description of nature culminating in a mystery of transmutation. It is the familiar alchemical process that by means of various chemical procedures releases from the *prima materia* the world-soul dormant in it and in so doing both redeems matter and transforms the adept. Fludd, unlike Kepler, had no original ideas of his own to proclaim; even his alchemical notions are formulated in a very primitive form. The universe is divided into four spheres, corresponding to the ancient doctrine of the four elements. The highest is the empyrean, the world of spirits, followed in descending order by the ether as the link with the sphere of the elements and sublunary things, and, at the bottom, by the earth, which is also the seat of the devil. The world is the mirror image of the invisible Trinitarian God who reveals Himself in it. Just as God is symbolically represented by an equilateral triangle so there is a second, reflected tri-

[31] Frisch, V, pp. 328–34.

[32] Frankfort on the Main, 1621 (called the *Discursus analyticus*).

[33] Frisch, V, pp. 413–68.

[34] Frankfort on the Main, 1622.

angle below that represents the world. This can be clearly seen in a figure taken from Fludd's *Utriusque Cosmi etc.* (Plate I).

Beside the upper triangle is the explanation (I):

That most divine and most beautiful Object [God] seen in the murky mirror of the world drawn underneath.

Divinissimum et formosissimum Illud objectum in subscripto aqueo speculo mundano conspectum.

Referring to the lower triangle (II):

The shadow, simulacrum, or reflection of the incomprehensible triangle seen in the mirror of the world.

Trianguli incomprehensibilis umbra simulacrum seu reflexio in speculo mundano visa.

In the upper triangle, Hebrew characters to be translated "Jahve" (?). In the text below we read (III):

Yet in so far as Hermes Trismegistus called the world the image of God Himself, in so far—I maintain—can the image and simulacrum of God Himself be discerned in the spirit of the world like the reflection of a man in a mirror.

At vero quatenus Trismegistus appellavit mundum ipsius Dei imaginem, eatenus ipsius Dei imaginem et simulacrum in mundi spiritu, tamquam effigiem humanam in speculo, conspici dicimus.

The two polar fundamental principles of the universe are *form* as the light principle, coming from above, and *matter* as the dark principle, dwelling in the earth. All beings from angels to minerals are differentiated only according to their greater or lesser light content. A constant struggle goes on between these polar opposites: from below, the material pyramid grows upward from the earth like a tree, the matter becoming finer toward the top; at the same time the formal pyramid grows downward with

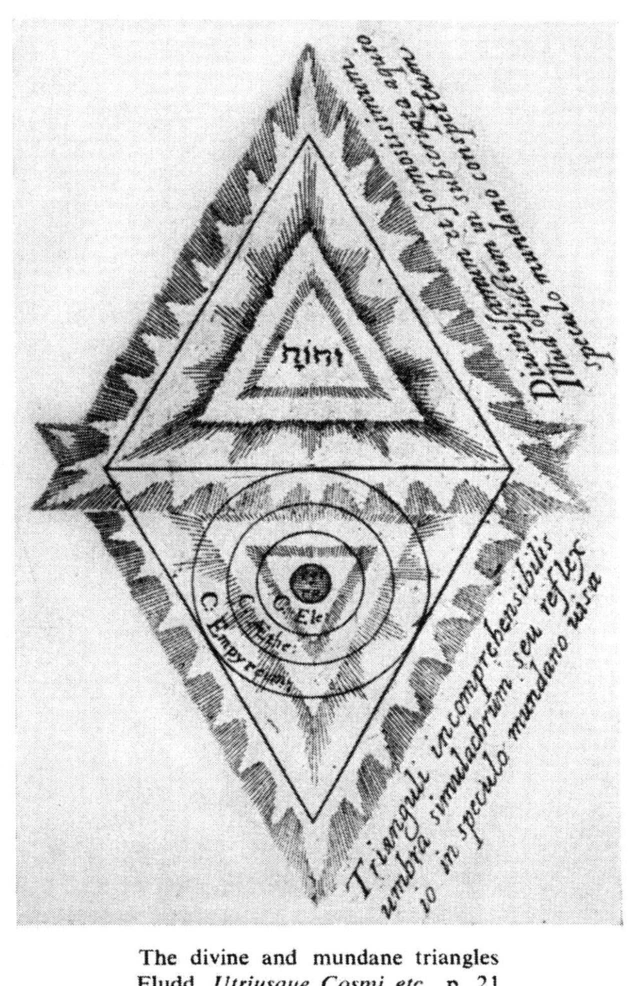

The divine and mundane triangles
Fludd, *Utriusque Cosmi etc.*, p. 21

PLATE I

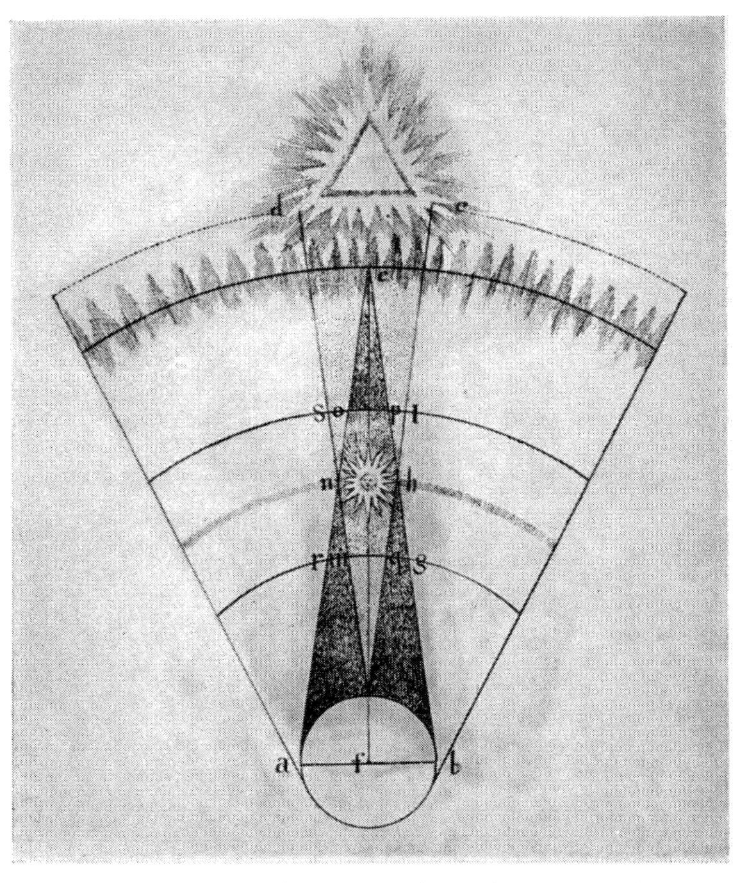

The interpenetration of the material and formal pyramids: 1
(With the *infans solaris*)
Fludd, *Utriusque Cosmi etc.*, p. 81

PLATE II

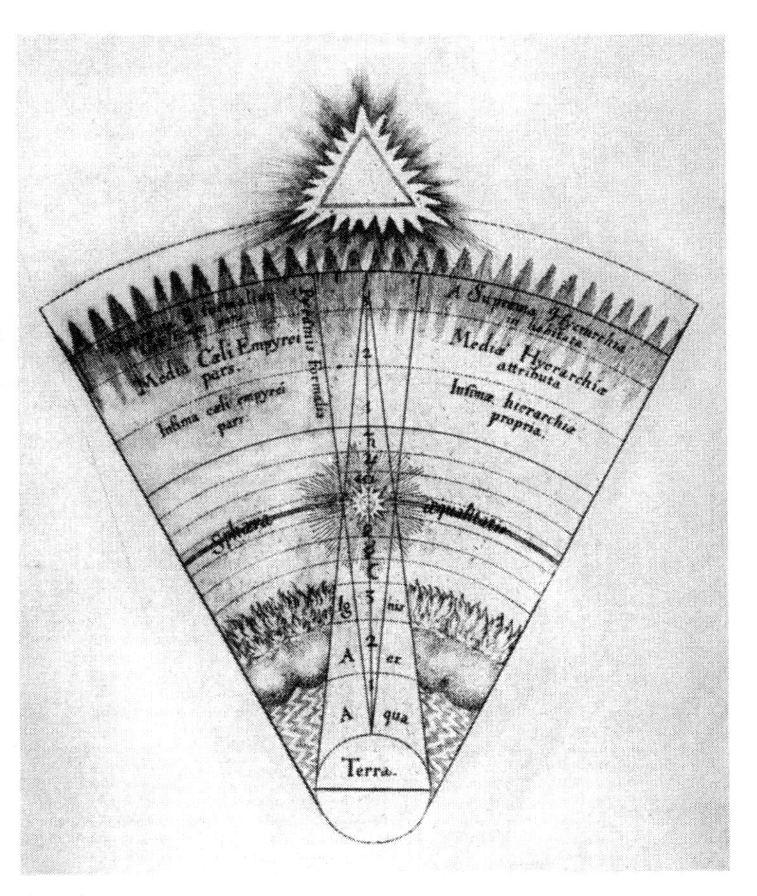

The interpenetration of the material and formal pyramids: 2
(With the *infans solaris*)
Fludd, *Utriusque Cosmi etc.*, p. 89

PLATE III

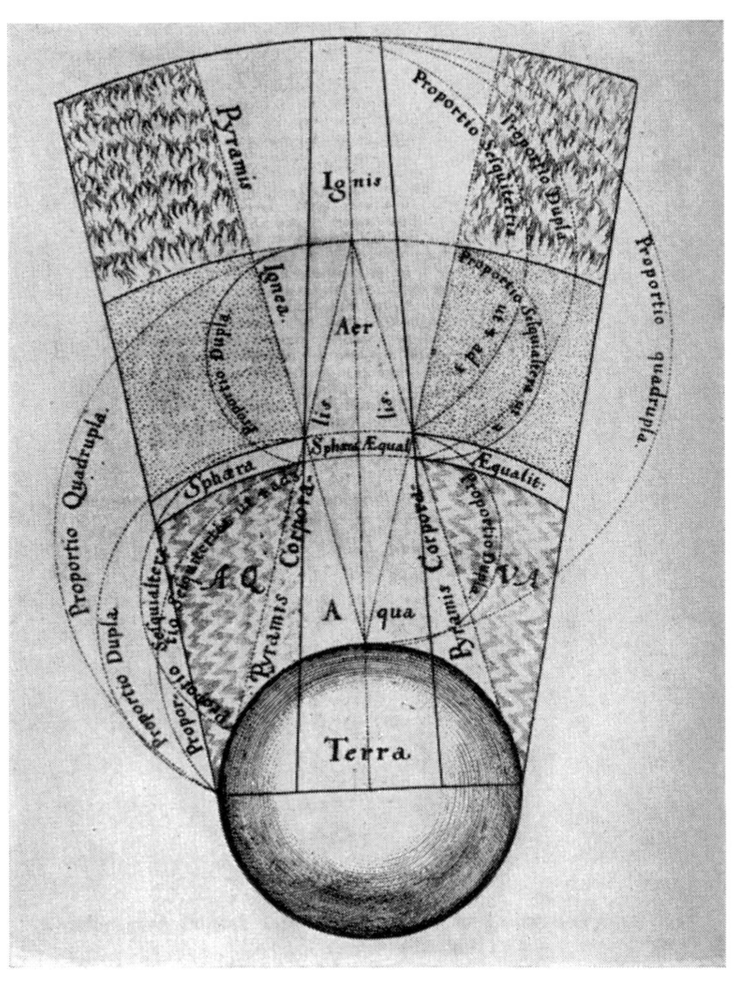

The interpenetration of the material and formal pyramids: 3
Fludd, *Utriusque Cosmi etc.*, p. 97

PLATE IV

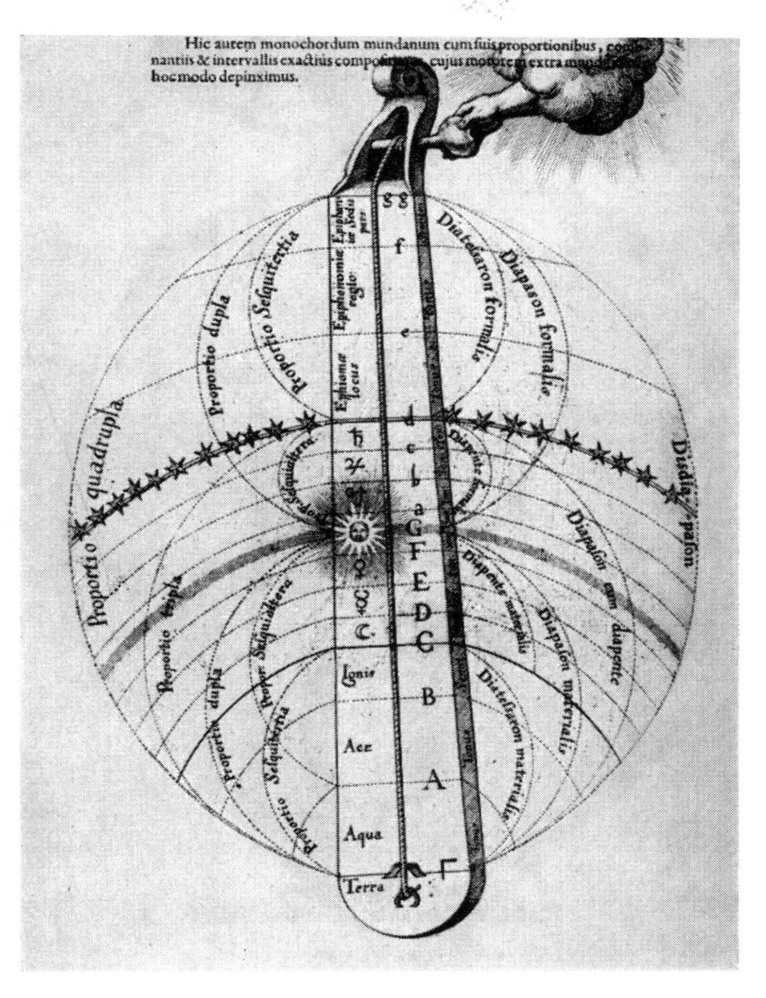

Monochordus mundanus
Fludd, *Utriusque Cosmi etc.*, p. 90

PLATE V

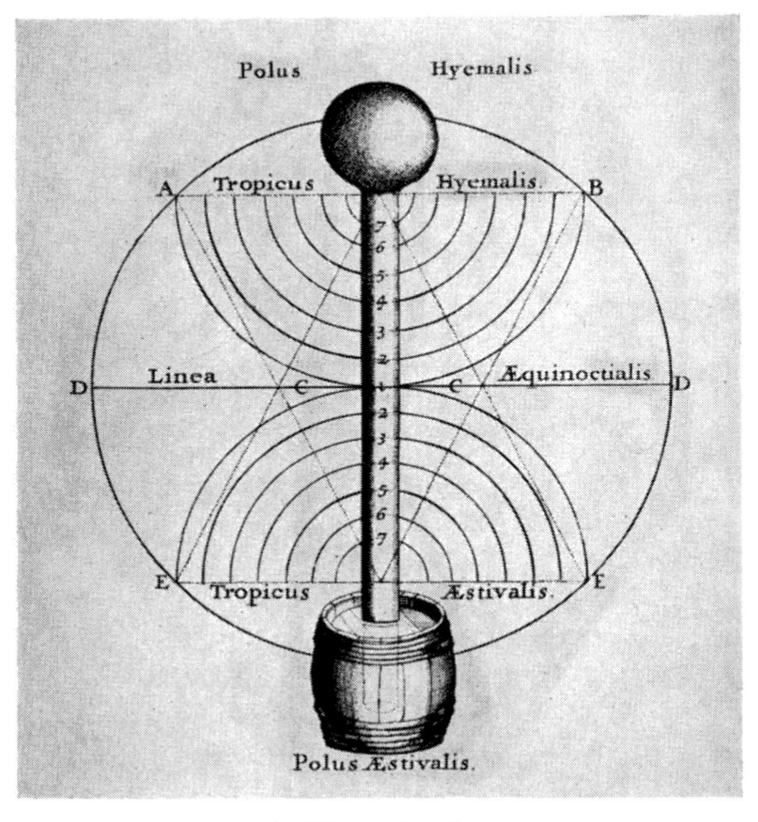

Fludd's weather-glass
Fludd, *Philosophia Moysaica*, fol. 4

PLATE VI

its apex on the earth, exactly mirroring the material pyramid. Fludd never distinguishes clearly between a real, material process and a symbolical representation. Because of the analogy of the microcosm to the macrocosm the chemical process is indeed at the same time a reflection of the whole universe. The two movements, the one downward and the other upward, are also termed sympathy and antipathy or, with reference to the cabala, *voluntas Dei* and *noluntas Dei.* After the withdrawal of the formal light principle matter remains behind as the dark principle, though it was latently present before as a part of God.[35] In the middle, the sphere of the sun, where these opposing principles just counterbalance each other, there is engendered in the mystery of the chymic wedding the *infans solaris,* which is at the same time the liberated world-soul. This process is described in a series of pictures (*picturae*) which Fludd also designates as "hieroglyphic figures" or "aenigmata." Plates II to IV serve as examples of this.

In agreement with old Pythagorean ideas Fludd evolves from the proportions of the parts of these pyramids the cosmic music, in which the following simple musical intervals play the chief part.

[35] This agrees with the *Tsimtsun* (Withdrawal) theory of the Cabalist Isaac Luria (1539–72; lived in Safad, Palestine). Cf. Gershom Scholem, *Major Trends in Jewish Mysticism* (New York, 1946), the seventh lecture. It seems to me that this mystical doctrine must be regarded as one of the attempts to harmonize the Aristotelian and alchemical idea of the *increatum* of the *prima materia,* which Fludd also accepts in its essentials, with the Biblical doctrine. The idea that matter existed from all eternity is also specifically advocated by the Italian philosopher Giacomo Zabarella (1532–89).

Disdiapason	= Double octave	*Proportio quadrupla*	4 : 1
Diapason	= Octave	*Proportio dupla*	2 : 1
Diapente	= Fifth	*Proportio sesquialtera*	3 : 2
Diatessaron	= Fourth	*Proportio sesquitertia*	4 : 3

This is expressed in the characteristic figure in Plate V, representing the *monochordus mundanus.* It may be remarked that the idea of cosmic music also appears in the works of the alchemist Michael Maier.

Fludd's general standpoint is that true understanding of world harmony and thus also true astronomy are impossible without a knowledge of the alchemical or Rosicrucian mysteries. Whatever is produced without knowledge of these mysteries is an arbitrary, subjective fiction. According to Kepler, on the other hand, only that which is capable of quantitative, mathematical proof belongs to objective science, the rest is personal. It is already apparent from the concluding words of the appendix to Book V of the *Harmonices mundi* [36] that Kepler had to fight in order to justify the adoption of strict mathematical methods of proof:

From this brief discussion I think it is clear that, although a knowledge of the harmonious proportions is very necessary in order to understand the dense mysteries of the exceedingly profound philosophy that Robert [Fludd] teaches, nevertheless the latter, who has even studied my whole work, will, for the time being, remain no less far removed from those perplexing

Ex his paucis constare arbitror, etsi ad intelligenda mysteria conferta philosophiae profundissimae, quam tradit Robertus, cognitione proportionum harmonicarum omnino opus est, tamen illum, qui vel totum opus meum edidicit, adhuc a mysteriis illis perplexissimis abfuturum haud paulo longius ac ipsae ab accuratissima certitudine demonstrationum mathe-

[36] Frisch, V, p. 334.

mysteries than these [the proportions] have receded [for him] from the accurate certainty of mathematical demonstrations.

maticarum recesserunt.

Fludd's aversion to all quantitative mensuration is revealed in the following passages:

What he [Kepler] has expressed in many words and long discussion I have compressed into a few words and explained by means of hieroglyphic and exceedingly significant figures, not, to be sure, for the reason that I delight in pictures (as he says elsewhere) but because I (as one of whom he seems to hint further below that he associates with alchemists and Hermetic philosophers) had resolved to bring together much in little and, in the fashion of the alchemists, to collect the extracted essence, to reject the sedimentary substance, and to pour what is good into its proper vessel; so that, the mystery of science having been revealed, that which is hidden may become manifest; and that the inner nature of the thing, after the outer vestments have been stripped off, may be enclosed, as a precious gem set in a gold ring, in a figure best suited to its nature—a figure, that is, in which its essence can be beheld by eye and mind as

Quod igitur ille multis verbis et longa oratione expressit, hoc ego brevibus contraxi, figurisque hieroglyphis et valde significantibus explicavi; non sane ideo, quia picturis delector (ut ipse alibi dicit) sed quoniam multa paucis congregare et more Chymicorum (quippe quem cum Chymicis et Hermeticis versari infra innuere videtur) extractam essentiam colligere, faeculentam vero substantiam reiicere, et quod bonum est in suo proprio vasculo collocare decreveram, ut detecto sic scientiae arcano occultum manifestaretur, reique natura interna exutis vestibus, more gemmae pretiosae aureo annulo insertae, figurae naturae suae magis aptae includeretur, in qua eius virtus, tanquam in speculo, absque verborum pluri-

in a mirror and without many-worded circumlocution.

morum circuitione oculo et animo conspiceretur.[37]

For it is for the vulgar mathematicians to concern themselves with quantitative shadows; the alchemists and Hermetic philosophers, however, comprehend the true core of the natural bodies.

Nam mathematicorum vulgarium est circa umbras quantitativas versari; Chymici et Hermetici veram corporum naturalium medullam amplectuntur.[38]

By the select mathematicians who have been schooled in formal mathematics nature is measured and revealed in the nude; for the spurious and blundering ones, however, she remains invisible and hidden. The latter, that is, measure the shadows instead of the substance and nourish themselves on unstable opinions; whereas the former, rejecting the shadow, grasp the substance and are gladdened by the sight of truth.

A Mathematicis exquisitis et circa mathesin formalem versatis mensuratur atque revelatur Natura nuda; a spuriis autem et mendosis invisibilis et occulta manet. Hi ergo umbras pro substantia metiuntur, opinionibus variis nutriuntur; illi, umbra rejecta, substantiam amplectuntur, veritatisque visione gaudent.[39]

But here lies hidden the whole difficulty, because he [Kepler] excogitates the exterior movements of the created thing [40] whereas I contemplate [41] the internal and essential impulses [42] that issue from nature herself;

Sed hic tota latet difficultas, quod ipse motus rei naturatae exteriores excogitat, ego actus internos et essentiales ab ipsa natura profluentes considero; ipse caudam tenet, ego caput amplector; ego causam princi-

[37] *Demonstratio analytica,* p. 5. [38] P. 12. [39] P. 13.

[40] *res naturata*: the actually existing natural object.

[41] Kepler "puzzles out," Fludd "beholds."

[42] The *actus interiores* are the creative impulses occurring in "nature herself" *(ipsa natura);* the *motus exteriores,* resulting from these impulses, are the physical events occurring in "the created things" *(res naturata).*

he has hold of the tail, I grasp the head; I perceive the first cause, he its effects. And even though his outermost movements may be (as he says) real, nevertheless he is stuck too fast in the filth and clay of the impossibility of his doctrine and, perplexed, is too firmly bound by hidden fetters to be able to free himself easily from those snares without damage to his honour and ransom himself from captivity cheaply.[43] And he who digs a pit for others will unwittingly fall into it himself.

palem, ipse illius effectus animadvertit. Et tamen ipse, quamvis motus eius extremi sint reales (ut dicit), magis coeno et luto impossibilitatis suae doctrinae inhaeret et perplexissimus obscuratis vinculis obligatur, quam, ut se facilis ex laqueis istis, salvo suo honore, liberare, captumque redimere queat minimo; atque qui foveam aliis fecit in eandem ipsemet ignoranter incidit.[44]

Such a rejection of everything quantitative in favour of the "forma" (we should say for "forma" symbol) is obviously completely incompatible with scientific thinking. Kepler replies to the above as follows:

When I pronounce your enigmas—harmonies, I should say—obscure I speak according to my judgment and understanding, and I have you yourself as an aid in this since you deny that your purpose is subject to mathematical demonstration without which I am like a blind man.

Quod igitur aenigmata tua, harmonica inquam, tenebrosa appello, loquor ex judicio et captu meo, et habeo te astipulatorem, qui negas, tuam intentionem subjici demonstrationibus mathematicis, sine quibus ego coecus sum.[45]

The disputants, then, can no longer even agree on what to call light and what dark. Fludd's symbolical *picturae* and Kepler's geometrical diagrams present an irrec-

[43] *minimo:* for a small price.

[44] Fludd, *Discursus analyticus,* p. 36.

[45] Frisch, V, p. 424.

oncilable contradiction. It is, for example, easy for Kepler to point out that the dimensions of the planetary spheres assumed in Fludd's figure of the *monochordus mundanus* illustrated above do not correspond to the true, empirical dimensions. When Fludd retorts that the *sapientes* are not agreed as to the ultimate dimensions of the spheres and that these are not essentially important, Kepler remarks very pertinently that the quantitative proportions are essential where music is concerned, especially in the case of the proportion 4 : 3, which is characteristic of the interval of the fourth. Kepler naturally objected, furthermore, to Fludd's assumption that the earth and not the sun is the centre of the planetary spheres.

Fludd's depreciation of everything quantitative, which in his opinion belongs, like all division and all multiplicity, to the dark principle (matter, devil),[46] resulted in a

[46] *Replicatio,* p. 27, on Franciscus Georges Venetus:

"He concludes therefore that the soul is one and simple, but can be called divisible when descending to the lower things. And this is the reason for generation and corruption in the lower spheres. For this reason, Pythagoras says, writing to Eusebius: 'God is in unity; but in duality is the Devil and evil, because in the latter is material multiplicity . . .' "

"Concludit igitur, quod anima sit unica et simplex, ad res vero inferiores descendens divisa dicitur. Atque haec est generationis et corruptionis ratio in rebus inferioribus. Hinc ergo dicit Pythagoras scribendo ad Eusebium: Deus est in unitate, in dualitate vero est Diabolus et malum quippe in quo est multitudo materialis . . ."

Replicatio, p. 37:

"Matter, expanding in multitude and not in form, which latter is always uninterruptedly connected with its shining source . . ."

"Materiam, quae sola in multitudine dilatatur et non in forma quae semper continua est ad suum fontem lucidum . . ."

further essential difference between Fludd's and Kepler's views concerning the position of the soul in nature. The sensitivity of the soul to proportions, so essential according to Kepler, is in Fludd's opinion only the result of its entanglement in the (dark) corporeal world, whereas its imaginative faculties, that recognize unity, spring from its true nature originating in the light principle (*forma*). While Kepler represents the modern point of view that the soul is a part of nature, Fludd even protests against the application of the concept "part" to the human soul, since the soul, being freed from the laws of the physical world, that is, in so far as it belongs to the light principle, is inseparable from the *whole* world-soul (see Appendix I).

Kepler is obliged to reject the "formal mathematics" that Fludd opposes to "vulgar" mathematics:

If you know of another mathematics (besides that vulgar one from which all those hitherto celebrated as mathematicians have received their name), that is, a mathematics that is both natural and formal, I must confess that I have never tasted of it, unless we take refuge in the most general origin of the word [teaching, doctrine] and give up the quantities. Of that, you must know, I do not speak here. You, Robert, may keep for yourself its glory and that of the proofs to be found in it —and how accurate and how certain those are, that, I think, you will judge for yourself

Mathesin si tu aliam nosti (praeter vulgarem illam, a qua denominati fuerunt quotquot hactenus mathematici celebrantur), quae scilicet sit naturalis et formalis, eam ego fateor numquam delibasse, nisi ad generalissimam vocis originem confugimus, dimissis quantitatibus. De illa igitur scito me hic non esse locutum; habeas tibi, Roberte, laudem et illius et demonstrationem in illa, quae quam sint accuratae, quam certae, tute tecum judicabis sine me arbitro.

without me. *I* reflect on the visible movements determinable by the senses themselves, *you may consider the inner impulses* and endeavour to distinguish them according to grades. *I hold the tail* but I hold it in my hand; *you may grasp the head mentally,* though only, I fear, in your dreams. I am content with the effects, that is, the movements of the planets. If you shall have found in the very causes harmonies as limpid as are mine in the movements, then it will be proper for me to congratulate you on your gift of invention and myself on my gift of observation —that is, as soon as I shall be able to observe anything.

Motus ego cogito visibiles sensuque ipso determinabiles, tu *actus internos considerato* deque iis in gradus distinguendis laborato; *caudam ego teneo* sed manu, tu *caput amplectaris* mente, modo ne somnians; ego contentus sum *effectis* seu planetarum motibus, tu si in ipsis causis invenisti harmonias adeo liquidas, quam sunt meae in motibus, aequum erit, ut ego et tibi de inventione et mihi de perceptione gratuler, ubi primum percipere potero.[47]

The situation is, however, not so simple as Kepler here represents it to be. His theoretical standpoint is, after all, not purely empirical but contains elements as essentially speculative as the notion that the physical world is the realization of pre-existent archetypal images. It is interesting that this speculative side of Kepler (here not avowed) is matched by a less obvious empirical tendency in Fludd. The latter tried in fact to support his speculative philosophy of the light and dark principle by means of scientific experiments with the so-called "weather-glass." Since this attempt casts light on what seems to us an extremely bizarre episode in the intellectual history of the seventeenth century, I should like to say something more about

[47] *Apologia* (Frisch, V, p. 460).

it at this point, although the relevant passages are only to be found in a late work of Fludd's, the *Philosophia Moysaica* (Gouda, 1637), which did not appear until after Kepler's death.

The weather-glass was constructed by immersing a glass vessel, opening downward, into a receptacle filled with water. The air contained in the vessel having been rarefied by heating, a column of water will rise within, its level determined by both temperature and air pressure. The latter concept, however, was not known before Torricelli, and the temporary variations in the water level, caused in part by variations in the air pressure, were usually interpreted as resulting from only the variations in temperature. On being warmed the column of water falls, on being cooled it rises, as a result of the expansion or contraction of the air remaining above the column of water. The instrument, a kind of combined thermometer and barometer, behaves, of course, in a way opposite of what we are used to.[48]

Plate VI and the following quotations from the *Philosophia Moysaica* make it apparent how Fludd regards the weather-glass as a symbol of the struggle between the light and dark principles in the macrocosm, a subject which has already been discussed here. The triangles on

[48] On the history of this instrument cf. G. Boffito, *Gli strumenti della scienza e la scienza degli strumenti* (Florence, 1929), where reference is made (Pl. 66) to an illustration and description of the "weather-glass" in Giuseppe Biancani's *Sphaera Mundi* (Bologna, 1620), p. 111, and also to Galileo's similar instrument called a "thermoscope" (Pl. 115).

I owe this bibliographical information to the kindness of Professor Panofsky.

Plate VI are the same as those in the earlier figures (Plates I–IV).

That the instrument commonly called a weather-glass is falsely arrogated unto themselves by some contemporaries; that is, that they falsely boast of it as being their own invention.

So zealously avid of renown and greedy for fame and reputation is man that how and by what right he acquires them, still less whether by straight or crooked means, is of small concern to him. This alone was the reason why the heathen philosophers ascribed to themselves fraudulently those philosophical principles that by highest right belonged to the wise and divine philosopher Moses, and veiled and, as it were, gilded this theft of theirs by means of new names or titles, so that in this fashion they could show them off as having been due to their own invention (as will be enlarged upon below). In a quite similar way this experimental instrument or glass of ours has many spurious or illegitimate inventors who, by altering the form of the original somewhat, boast that they have first thought of this idea (*inventionem*) by themselves. As far as I am concerned, I judge it to be just and

Quod instrumentum vulgo speculum Calendarium dictum, falso a quibusdam nostri seculi hominibus sibimet ipsis arrogatur, utpote, qui illud propriam suam inventionem esse falso gloriantur.

Gloriae tam impense avidus atque famae et reputationis cupidus est homo, ut quomodo, quave ratione illam acquirat, nimirum an sit directe vel indirecte, parum refert. Ista sola erat causa, ob quam Ethnici philosophi sibi ipsis ea philosophiae principia more surreptitio ascripserunt, quae summo jure sapienti divinoque philosopho Moysi pertinebant, nominibusque sive titulis novis illam suam latrociniam velabant et quasi deaurabant, ut hac ratione ostentarent ea propriis suis inventionibus fuisse stabilita (ut infra dicetur latius); simili plane ratione instrumentum sive speculum hoc nostrum experimentale, plurimos habet inventores spurios seu adulterinos, qui, quoniam typi formam aliquantulum immutarunt, ipsius inventionem a seipsis prius excogitatam gloriantur. Quod ad me attinet, cuilibet quod suum erit tribuere aequam atque honestum esse existimo: non enim erit mihi dedecus

honest to attribute to anyone what is his: for it is no shame to me to ascribe the principles of my philosophy to my teacher Moses who received them himself formed and written down by the finger of God. And therefore I cannot in justice arrogate to myself, or claim, the invention of this instrument, although I have made use of it (albeit in other form) in my history of the natural macrocosm and elsewhere in order to prove the truth of my philosophical argument. And I confess that I found it verbally specified and geometrically delineated in a manuscript at least fifty years old. First then, I shall explain to you the form in which I found it in that old record just mentioned; then I shall describe its shape and position as it is commonly known and used among us.

istius meae philosophiae principia praeceptori meo Moysi ascribere, utpote qui ipsa etiam divino digito formata atque designata accepit, neque jure mihi fabricam huius instrumenti primariam arrogare aut vendicare queam, quamvis illo in naturali Macrocosmi, mei historia et alibi ad veritatem argumenti mei philosophici demonstrandam (licet in alia forma) sum usus: et agnosco, me illud in veteri quingentorum saltem annorum antiquitatis manuscripto graphice specificatum, atque geometrice delineatum invenisse. Primo itaque formam, sub qua illud in monumento praedicto antiquo inveni, vobis exponam: deinde eius figuram atque positionem, quod vulgariter inter nos est cognitum atque usitatum hic describam.[49]

Before we proceed to our ocular demonstration, which will be performed by means of our experimental instrument, we ought first to consider that the general air, that is, the general element of the sublunary world, is the thinner and more spiritual portion of the "waters beneath the firmament" of which Moses makes mention. There-

Priusquam ad ocularem nostram istam demonstrationem procedamus, quae erit in et per experimentale nostrum instrumentum facta, inprimis considerare debemus, quod Catholicus aer seu generale regionis sublunaris Elementum, sit subtilior et magis spiritualis Aquarum, infra Firmamentum portio, de quibus Moyses facit

[49] *Philosophia Moysaica,* I, 1, 2 fol. altera.

fore it is certain that any part of this air corresponds to its whole, and consequently the air enclosed in the glass of this Instrument is of the same nature and condition as the general world air. Wherefore it is clear that, because of the continuity between the two, the general air of the sublunary world behaves in its disposition exactly like the partial air enclosed in the glass which is a part of the general air; and this in turn behaves like the Spirit Ruach-Elohim that hovered above the waters, animated them by His presence, enlivened and informed them, and expanded them by giving them motion; so that in His absence, that is, upon the cessation of His actuating force and active emanation or upon the contraction of the activity of His rays back into Himself, the waters then correspondingly contracted, condensed, darkened, and were rendered motionless and quiet.

mentionem: Quare certum est quod quaelibet eiusdem aeris particula correspondeat eius toti, et per consequens aer inclusis in Instrumenti huius vitro est eiusdem naturae et conditionis cum aere Catholico mundano. Unde liquet, quod ratione continuitatis ipsorum, ut aer generalis mundi sublunaris in sua dispositione se habet, ita etiam eius aer particularis vitro inclusa, qui est Catholici pars, se habet iterum ut Spiritus Ruah-Elohim, qui ferebatur super aquas, ipsas sua praesentia animavit, vivificavit, informavit, easque dando iis motionem dilatavit; ita quidem ipsius absentia seu actus et emanationis agilis cessatione, seu radiorum activitatis suae in seipsum contractione, aquae similiter sunt contractae, condensatae, obscuratae et immobiles atque quietae factae.[50]

In view of this description one would almost be inclined to call the weather-glass in Fludd's sense a "noluntometer."

It is significant for the psychological contrast between Kepler and Fludd that for Fludd the number four has a special symbolical character, which, as we have seen, is not true of Kepler. A quotation from Fludd's *Dis-*

[50] Ibid., fol. 27 v. (I, III).

cursus analyticus, given in Appendix II, will throw some light on this matter.

From what has been said above the reader has gained, we hope, some understanding of the prevailing atmosphere of the first half of the seventeenth century when the then new, quantitative, scientifically mathematical way of thinking collided with the alchemical tradition expressed in qualitative, symbolical pictures: the former represented by the productive, creative Kepler always struggling for new modes of expression, the latter by the epigone Fludd who could not help but feel clearly the threat to his world of mysteries, already become archaic, from the new alliance of empirical induction with mathematically logical thought. One has the impression that Fludd was always in the wrong when he let himself be drawn into a discussion concerning astronomy or physics. As a consequence of his rejection of the quantitative element he perforce remained unconscious of its laws and inevitably came into irreconcilable conflict with scientific thinking.

Fludd's attitude, however, seems to us somewhat easier to understand when it is viewed in the perspective of a more general differentiation between two types of mind, a differentiation that can be traced throughout history, the one type considering the quantitative relations of the *parts* to be essential, the other the qualitative indivisibility of the *whole.* We already find this contrast, for example, in antiquity in the two corresponding definitions of beauty: in the one it is the proper agreement of the parts with each other and with the whole, in the other (going back to Plotinus) there is no reference to parts but beauty is the eternal radiance of the "One" shining through the material phenomenon.[51] An analogous contrast can also be found later in the well-known quarrel between Goethe and Newton concerning the theory of colours: Goethe had a similar aversion to "parts" and always emphasized the disturbing influence of instruments on the "natural" phenomena. We should like to advocate the point of view that these controversial attitudes are really illustrations of the psychological contrast between feeling type or intuitive type and thinking type. Goethe and Fludd represent the feeling type and the intuitive approach, Newton and Kepler the thinking type; even Plotinus should probably not be called a systematic thinker, in contrast to Aristotle and Plato.[52]

Just because the modern scholar prefers in principle not to ascribe to either one of these two opposite types a higher degree of consciousness than to the other, the old historical dispute between Kepler and Fludd may still be considered interesting as a matter of principle even in an age for which both Fludd's and Kepler's scientific ideas about world music have lost all significance. An added indication of this can be seen in particular in the fact that the "quaternary" attitude of Fludd corresponds, in contrast to Kepler's "trinitarian" attitude, from a psychologi-

[51] The controversy between these two definitions of beauty plays an important role particularly in the Renaissance, where Ficino took his stand entirely on the side of Plotinus.

[52] In so far as scientific thought, based on the co-operation of experiment and theory, is a combination of thinking and sensation, its opposite pole can be more precisely expressed by the term "intuitive feeling." On Plotinus cf. also Schopenhauer, *Fragmente zur Geschichte der Philosophie,* 7: "Neuplatoniker" (in the *Parerga und Paralipomena,* ed. by R. von Koeber, Berlin, 1891).

cal point of view, to a greater *completeness of experience.*[53] Whereas Kepler conceives of the soul almost as a mathematically describable system of resonators, it has always been the symbolical image that has tried to express, in addition, the immeasurable side of experience which also includes the imponderables of the emotions and emotional evaluations. Even though at the cost of consciousness of the quantitative side of nature and its laws, Fludd's "hieroglyphic" figures do try to preserve a *unity* of the inner experience of the "observer" (as we should say today) and the external processes of nature, and thus a *wholeness* in its contemplation—a wholeness formerly contained in the idea of the analogy between microcosm and macrocosm but apparently already lacking in Kepler and lost in the world view of classical natural science.[54]

Modern quantum physics again stresses the factor of the disturbance of phenomena through measurement (see the following section), and modern psychology again utilizes symbolical images as raw material (especially those that have originated spontaneously in dreams and fantasies) in order to recognize processes in the collective ("objective") psyche. Thus physics and psychology reflect again for modern man the old contrast between the

[53] This is in harmony with the older alchemical texts according to which only the totality of all four elements makes it possible to produce the *quinta essentia* and the *lapis,* that is, the actual transmutation. Further remarks on the symbolism of the numbers three and four will be found in Appendix III.

[54] As modern parallels to this tendency toward unity and wholeness, cf. especially Jung's study of synchronicity in this volume, and his essay, "The Spirit of Psychology," in *Spirit and Nature* (Papers from the Eranos Yearbooks, 1; New York, 1954; London, 1955).

quantitative and the qualitative. Since the time of Kepler and Fludd, however, the possibility of bridging these antithetical poles has become less remote. On the one hand, the idea of complementarity in modern physics has demonstrated to us, in a new kind of synthesis, that the contradiction in the applications of old contrasting conceptions (such as particle and wave) is only apparent; on the other hand, the employability of old alchemical ideas in the psychology of Jung points to a deeper unity of psychical and physical occurrences. To us, unlike Kepler and Fludd, the only acceptable point of view appears to be the one that recognizes *both* sides of reality—the quantitative and the qualitative, the physical and the psychical—as compatible with each other, and can embrace them simultaneously.

7

It is obviously out of the question for modern man to revert to the archaistic point of view that paid the price of its unity and completeness by a naïve ignorance of nature. His strong desire for a greater unification of his world view, however, impels him to recognize the significance of the pre-scientific stage of knowledge for the development of scientific ideas—a significance of which mention has already been made at the beginning of this essay—by supplementing the investigation of this knowledge, directed inward. The former process is devoted to adjusting our knowledge to external objects; the latter should bring to light the archetypal images used in the creation of our scientific concepts. Only by combining both these directions of research may complete understanding be obtained.

Among scientists in particular, the universal desire for a greater unification of our world view is greatly intensified by the fact that, though we now have natural sciences, we no longer have a total scientific picture of the world. Since the discovery of the quantum of action, physics has gradually been forced to relinquish its proud claim to be able to understand, in principle, the *whole* world. This very circumstance, however, as a correction of earlier one-sidedness, could contain the germ of progress toward a unified conception of the entire cosmos of which the natural sciences are only a part.

I shall try to demonstrate this by reference to the still unsolved problem of the relationship between occurrences in the physical world and those in the soul, a problem that had already engaged Kepler's attention. After he had shown that the optical images on the retina are inverted in relation to the original objects he baffled the scientific world for a while by asking why then people did not see objects upside down instead of upright. It was of course easy to recognize this question as only an illusory problem, since man is in fact never able to compare images with real objects but only registers the sensory impressions that result from the stimulation of certain areas of the retina. The general problem of the relation between psyche and physis, between the inner and the outer, can, however, hardly be said to have been solved by the concept of "psychophysical parallelism" which was advanced in the last century. Yet modern science may have brought us closer to a more satisfying conception of this relationship by setting up, within the field of physics, the concept of *complementarity*. It would be most satisfactory of all if physis and psyche could be seen as complementary aspects of the same reality. We do not yet know, however, whether or not we are here confronted—as surmised by Bohr and other scientists—with a true complementary relation, involving mutual exclusion, in the sense that an exact observation of the physiological processes would result in such an interference with the psychical processes that the latter would become downright inaccessible to observation. It is, however, certain that modern physics has generalized the old confrontation of the apprehending subject with the apprehended object into the idea of a cleavage or division that exists between the observer or the means of observation, on the one hand, and the system observed, on the other. While the *existence* of such a division is a necessary condition of human cognition, modern physics holds that its *placement* is, to a certain extent, arbitrary and results from a choice co-determined by considerations of expediency and hence partially free. Furthermore, whereas older philosophical systems have located the psychical on the subjective side of the division, that is, on the side of the apprehending subject, and the material on the other side—the side of that which is objectively observed—the modern point of view is more liberal in this respect: microphysics shows that the means of observation can also consist of apparatuses that register automatically; modern psychology proves that there is on the side of that which is observed introspectively an unconscious psyche of considerable objective reality. Thereby the presumed objective order of nature is, on the one hand, relativized with respect to the no less indispensable means of observation outside the observed system; and, on the other, placed beyond the distinction of "physical" and "psychical."

Now, there is a basic difference between the observers, or instruments of observation, which must be taken into consideration by modern microphysics, and the detached observer of classical physics. By the latter I mean one who is not necessarily without effect on the system observed but whose influence can always be eliminated by determinable corrections. In microphysics, however, the natural laws are of such a kind that every bit of knowledge gained from a measurement must be paid for by the loss of other, complementary items of knowledge. Every observation, therefore, interferes on an indeterminable scale both with the instruments of observation and with the system observed and interrupts the causal connection of the phenomena preceding it with those following it. This uncontrollable interaction between observer and system observed, taking place in every process of measurement, invalidates the deterministic conception of the phenomena assumed in classical physics: the series of events taking place according to pre-determined rules is interrupted, after a free choice has been made by the beholder between mutually exclusive experimental arrangements, by the selective observation which, as an essentially non-automatic occurrence, may be compared to a creation in the microcosm or even to a transmutation the results of which are, however, unpredictable and beyond human control.[55]

[55] Cf. on this matter the author's essay "Die philosophische Bedeutung der Idee der Komplementarität" in *Experientia* (Basel),

In this way the role of the observer in modern physics is satisfactorily accounted for. The reaction of the knowledge gained on the gainer of that knowledge gives rise, however, to a situation transcending natural science, since it is necessary for the sake of the completeness of the experience connected therewith that it should have an obligatory force for the researcher. We have seen how not only alchemy but the heliocentric idea furnishes an instructive example of the problem as to how the process of knowing is connected with the religious experience of transmutation undergone by him who acquires knowledge. This connection can only be comprehended through symbols which both imaginatively express the emotional aspect of the experience and stand in vital relationship to the sum total of contemporary knowledge and the actual process of cognition. Just because in our times the possibility of such symbolism has become an alien idea, it may be considered especially interesting to examine another age to which the concepts of what is now called classical scientific mechanics were foreign but which permits us to prove the existence of a symbol that had, simultaneously, a religious and a scientific function.

VI (1950) : 2, 72. The new type of statistical, quantum-physical natural law, which functions as an intermediary between discontinuum and continuum, cannot in principle be reduced to causal-deterministic laws in the sense of classical physics; and just in limiting that which happens according to law to that which is reproducible it must recognize the existence of the essentially unique in physical occurrences. I should like to propose, following Bohr, the designation "statistical correspondence" for this new form of natural law.

APPENDIX I

FLUDD'S REJECTION OF THE PROPOSITION THAT THE SOUL OF MAN IS A PART OF NATURE

Replicatio in Apolog. ad Anal. XII (Frankfort on the Main, 1622), pp. 20 f.[1]

From these foundations of your *Harmonices* there arise, it seems to me, multifarious questions and doubts not easy to resolve, namely:

1. Whether the human soul is a part of nature?

2. Whether the circle with its divisions by the regular polygons is reflected in the soul because it [the soul] is an image of God?

3. Whether the determinants of the intellectual harmonies in the Mind Divine are established on the basis of the division of the circle which takes place in the essence of the soul itself, as Johannes Kepler would have it

Ut mihi videtur, ex hisce Harmonicae tuae fundamentis quaestiones et dubia multifaria non facile dissolvenda oriuntur: videlicet

1. An anima humana sit pars naturae?

2. An in Anima reluceat Circulus cum suis divisionibus per regularia plana, propterea quia ipsa est imago Dei.

3. An ex divisione, quae sit in ipsius animae essentia, constituantur termini harmoniarum intellectualium in mente divina, ut vult Johannes Keplerus

[1] Cf. Kepler's *Apologia* (Frisch, V, p. 429).

(p. 21), whose model here is the human mind, which has retained from its archetype the impress of the geometrical data since the very beginning of man?

(p. 21) cuius exemplar est hic humana, characterem rerum Geometricarum inde ob ortu hominis ex Archetypo suo retinens?

4. Whether the sense of hearing is a part of nature and bears witness to the sounds and their qualities as represented [to the intellect] by the *sensus communis?*

4. An Auditus pars sit Naturae, testeturque de sonis eorumque qualitatibus, quas sensus communis repraesentat?

5. Assuming that (on the basis of the aforesaid) the proportion is reflected in the mind from its origin, whether then the sounds should be considered harmonious and whether pleasure can be derived from them or not?

5. Si inveniretur proportio (ex praedictis) in intellectu ab origine relucens, an soni censeantur harmonici et utrum ab iis delectatio oriatur, necne?

6. Whether the triangle is a part of the nature of the intelligible things, likewise the square and whatever else divides the circle into parts which by their quantity or length determine[2] any harmonious proportion, and whether all other natural factors that are present in artificial song, follow the numerical value, so established, of the consonances?

6. An pars naturae rerum intelligibilium sit triangulum, pars quadrangulum et quodlibet distinguat circulum in partes, quae sunt quantitate seu longitudine sua termini proportionis alicuius harmonicae, et an ad numerum consonantiarum sic constitutum sequantur reliqua, quae insunt in cantu artificiali Naturalia?

On the main points of these questions, my Johannes, I shall begin to speak in order, not intending to contradict you in any way or to do any damage to your *Harmonices* but only

De harum, inquam quaestionum praecipuis, mi Johannes, ordine, non ut tibi in re aliqua contradicam, aut aliquid Harmonicae tuae damni afferam, sed disputandi solummodo gra-

[2] Literally: "are the *termini*," viz., determinants.

for the sake of discussion and as a philosopher stimulated by another philosopher to solve some questions, quite apart from his own opinion:

Whether the human soul is a part of nature?

This question I must answer in the negative, contrary to what you hope.

1. Because nature, in its capacity of universal soul, contains the formula of the whole and is not even divisible into essential parts, as Plato testifies.

2. Hermes Trismegistus says that the soul, or the human mind (which he did not hesitate to call the nature of God), can as little be separated or divided from God as a sunbeam from the sun.

3. Plato as well as Aristotle seem to affirm that the Creator of all things possessed as soul something whole [total] before any division. And Plato called this soul universal nature.

4. Plato says that the soul, when separated from the corporeal laws, is not a number having a definite quantity and cannot be divided into parts or multiplied but is of *one* form [a continuum].

5. And Iamblichus seems to maintain that the soul, though it seems to have within itself all orders and categories, is nevertheless always determined

tia, atque ut Philosophus a Philosopho ad quaestionum quarundam resolutionem praeter opinionem suam irritatus, sic exordior:

An Anima humana sit pars Naturae?

Quaestio haec negative a me contra spem tuam tenetur:

1. Quia Natura quatenus anima universalis rationem habet totius, nec in partes quidem essentiales dividitur, ut testatur Plato.

2. Dixit Mercurius Trismegistus, Animam s. mentem humanam (quam Dei naturam appellare haud dubitavit), a Deo non minus separari aut dividi, quam radius Solis a Sole.

3. Plato cum Aristotele affirmare videtur, quod creator omnium possideret animam totale quiddam ante divisionem. Et Plato hanc animam Universalem naturam nuncupavit.

4. Plato dicit, quod Anima separata a legibus corporeis non sit numerus habens quantitatem, nec dividitur nec multiplicatur in partes, sed est uniformis.

5. Et Jamblichus adstipulari videtur, quod Anima, quamvis videatur omnes rationes et species in se habere, tamen determinata est semper secundum

according to some unity.

6. Finally, Pythagoras and all the other philosophers who were endowed with some touch of the divine recognized that God is one and indivisible. Wherefore we can argue syllogistically as follows:

A. *That which was a whole before any division is not a part of something.*

B. *Now, the soul was a whole before any division.*

C. *Therefore it cannot be a part of nature.*[3]

B is proved by the third axiom mentioned above. But if you say in objection to A that the Philosopher meant the world-soul or the universal soul, whereas you mean the human soul, I reply with the fourth axiom that the soul separated from the corporeal laws is not a number and not divisible. Now that world-soul, which in Plato's opinion according to axiom 3 is nature itself, is separated from the corporeal laws. Consequently the human soul can also not be considered a part of the former since it is indivisible (as is proved by axioms 2, 3, and 4). Or I can deal with you in another way,

aliquid unum.

6. Pythagoras denique et omnes alii Philosophi divinitate aliqua praediti Deum agnoverunt esse unum et indivisibile. Sic ergo Syllogistice disputamus:

Quod erat totale quiddam ante ullam divisionem, non est pars alicuius rei,

At Anima erat totale quid ante divisionem,

Ergo non potest esse pars naturae.

Minor probatur per tertium axioma supra allegatum. At si ad Maiorem dicis, Philosophum intellexisse de Anima mundi seu totali, te autem de illa humana, replicamus cum axiomate quarto, quod Anima separata a corporeis legibus non est numerus, neque dividitur: At Anima illa mundi, quae secundum Platonem iuxta axioma 3 est ipsa natura, separatur a corporeis legibus, ergo nec Anima humana potest recenseri pro parte illius, cum sit indivisibilis, ut per 2, 3 et 4. Vel aliter sic tecum agam, argu-

[3] I designate the parts of the syllogism by A, B, C. A is what Fludd later calls *maior*, the major premise, the more general statement; B is *minor*, the minor premise, the more specific statement; C is the conclusion.

by taking my argument from your own mouth:

A. *The image of God is not part of anything.*

B. *Now, on the basis of what has been granted, the human soul is the image of God.*

C. *Therefore, it is not a part of nature.*

A is clear because God is the One and Indivisible according to axiom 6. B is your own assertion as it is cited in the second question and as the speech of Hermes Trismegistus about the extent of the mind declares, according to axiom 2.

Now we shall go on to the second question: *whether the circle with its divisions by the regular polygons is reflected in the soul because the latter is the image of God?*

I shall not hesitate to answer this question also in the negative, supported by the strongest and most encouraging arguments of the Philosophers. Namely, because:

1. Plato, first of all, says that the soul, separated from the corporeal laws, is not a number having quantity and is neither divisible nor multiplicable. But it is uniform, revolving in itself, rational, and surpasses all corporeal and material things.

2. Aristotle and Plato say that the Creator maintained

mentum meum a tuo proprio ore desumendo:

Imago Dei non est pars alicuius rei,

At vero, ex concessis, Anima humana est imago Dei,

Ergo non est pars Naturae.

Maior patet, quia Deus est unum et indivisibile, per 6. Minor est assertio tua, ut in quaestione secunda declaratur et Trismegisti sermo de Mentis amplitudine hoc declarat. Axioma 2.

Jam vero ad secundam Quaestionem properabimus,

An in Anima reluceat Circulus cum suis divisionibus per regularia plana, propterea quia ipsa est imago Dei?

Hanc etiam Quaestionem validissimis Philosophorum suffragiis stipatus et ad hoc incitantibus negare non haesitabo. Videlicet quoniam

1. Imprimis Plato dicit, quod Anima separata a legibus corporeis non est numerus habens quantitatem, unde nec dividitur illa nec multiplicatur. Sed est uniformis, in se revertens, et rationabilis, quae superat omnes res corporeas et materiales.

2. Aristoteles ac Plato dicunt, quod Creator retinuerit ani-

the soul as a totality before any division, and Pythagoras makes it a "one in itself" and says that it has its unity in the intellect.

mam totale quiddam ante divisionem: et Pythagoras ipsam in se ipsa unum facit, dicitque illam unitatem suam habere in intellectu.

3. Pythagoras, in his letter to Eusebius,[4] acknowledges that God is a unity and indivisible and says that duality is the Devil and evil, because in it lies multiplicity and materiality. And Plato holds that all good exists as One; but evil, he holds, comes from chaotic multiplicity.

3. Pythagoras ad Eusebium agnoscit, Deum esse unitatem et indivisibilem dicitque dualitatem esse Diabolum et malum, quippe in qua est multitudo et materialitas. Et Plato vult omne bonum esse per unum: at Malum vult esse propter multitudinem confusam.

4. Cicero says that it would not be possible for perfect order to exist in all the parts of the world unless they were united[5] by one single divine and continuous spirit.

4. Cicero dicit, quod non possit esse ordinis perfectio in omnibus mundi partibus, nisi de uno solo divino et continuo spiritu non essent continuatae.

5. God can neither be limited [defined] nor divided nor composed (according to Franciscus Georgius).

5. Deus nec potest definiri, nec dividi nec componi. Franciscus Georgius.

6. By the Platonic philosophers God is said to be present [lit.: poured into] in all things. [He is called] the world-soul (which, they say, contains the formula of the whole) inasmuch as He, universally diffused, fills and invigorates all things.

6. Deus infusus in omnibus rebus a Platonicis dicitur. Anima mundi [scil., dicitur], quam dicunt rationem habere totius, quatenus universaliter diffusus implet et vigorat omnia.

7. God can be determined neither according to essence nor according to quality nor

7. Deus est necque quid, necque quale, nec quantum,

[4] Cf. the "letter" quoted above, p. 198, n. 46.

[5] *continuatae:* united, brought into continuous connection.

according to quantity, inasmuch as no predication can comprehend Him. Scotus.

quatenus eum nullum comprehendit praedicamentum. Scotus.

8. The Pythagoreans and the Platonists regard the world-soul as being enclosed within the seven planetary spheres and say that within the first sphere it rests in the highest mind; and then, they say, it has become identical with it.

8. Pythagorici et Platonici includentes Animam inter 7. limites, dicunt ipsam in primo limite quiescere in summo intellectu et tum dicunt ipsam factam esse idem cum eo.

9. As all numbers are in the One, as all radii of the circle are in the centre, and as the powers of all the members are in the soul, so, it is said, is God in all things and all things in God. *Ars chymica.*

9. Sicut omnes numeri sunt in unitate, sicut in centro sunt omnes lineae circuli, sicut membrorum vires sunt in anima, sic Deus dicitur in omnibus et omnia in Deo. Ars Chym.

10. Hermes Trismegistus says that God is the centre of any one thing—a centre the periphery of which is nowhere.[6]

10. Mercurius Trismegistos dicit, quod Deus est cuiuslibet rei centrum, cuius circumferentia est nullibi.

With the help of these axioms of the Philosophers I therefore argue thus against your assertions.

Sic igitur super ista Philosophorum Axiomata contra vestram assertionem argumentamur.

ARGUMENT I.

A. *That which in and of itself is neither a number nor has quantity*[7] *is not capable of receiving into itself any quantitative* [*measurable*] *figure (such as the circle).*

Arg. I

Quod per se sumptum non est numerus nec quantitatem habet illud quidem figuram quantitativam (qualis est circulus) in se non recipit,

[6] A statement repeated many times by Fludd. This quotation comes from St. Bonaventura, *In Sententias,* I, d. 37, pars 1, a. 1, q. 1; but parallels abound in medieval literature. The source, it appears, is the pseudo-Hermetic *Liber XXIV Philosophorum* (12th cent.; see D. Mahnke, *Unendliche Sphäre und Allmittelpunkt,* Halle, 1937).

[7] Measurable size.

B. *Now the soul, which is freed from corporeal laws, is not a number and has no quantity.*

C. *Therefore the soul does not receive into itself from the very beginning a measurable figure (such as the circle); and consequently a circle is not in the least reflected in it.*

A is clear because a non-quantum[8] cannot receive into itself any quantities, as the One does not admit of multiplicity, and consequently is not a number.

B is confirmed by the first axiom and similarly by the second and third, according to which it is proved that the soul is one. But if you reply that the soul, as you conceive it, is not separated from physical laws since it is the human soul, then I say that you have meant the essence of the soul, as is apparent from your subsequent words; and in man, as he exists, this essence is not different from that soul of the macrocosm of which axiom 1 speaks in the second question, and according to axiom 2 of the first question where it is shown that the essence of the soul cannot be separated from God.

Or also thus:

A. *If the soul is an image of*

At Anima separata a legibus corporeis non est numerus, nec quantitatem habet,

Ergo anima figuram quantitativam (qualis est circulus) in se ab origine non recipit, et per consequens, circulus in ea minime relucet.

Maior patet in eo, quod non quantum quantitates recipere non potest, quemadmodum unum non admittit multitudinem, et consequenter, non est numerus. Minor confirmatur per axioma 1 et similiter per 2 et 3, quibus Anima profatur esse unum. Quod si respondeas, Animam, quatenus a te accipitur, non esse a legibus separatam, quippe humanam; dico ego, te de animae essentia intellexisse, quemadmodum ex sequentibus apparet, quae in homine existente non differt ab illa magni mundi, de qua intendit Axioma 1. in secunda quaestione, et per 2. primae, ubi probatur animae essentiam non posse dividi a Deo. Vel aliter sic:

Si anima sit imago Dei, non

[8] A non-quantitative magnitude.

God it is neither a quantity nor a number.

est quantitas nec numerus,

B. *Now it is, as you yourself admit, the image of God.*

At, te confitente, est Imago Dei,

C. *Therefore it is not a number nor does it admit of quantity.*

Ergo nec numerus est, nec quantitatem admittit.

A is established because God, according to axiom 7, cannot be determined according to essence nor according to quality or quantity, inasmuch as he stands outside of and above any predication.

Maior constat, quia Deus est nec quid, nec quale, nec quantum per axioma 7, quatenus extra et supra omne praedicamentum.

As far as the confirmation of your statement (demonstrating that the soul is the image of God) is concerned, however, this is also proved by axiom 7, which testifies to the fact that the soul rests at all times in God and becomes one with Him in the highest terminating sphere of its being. And [it is also proved] by axiom 2 in the first question according to which the mind is not divided from God.

Quod autem ad confirmationem tuae sententiae (Animam Dei imaginem probanti) attinet, illud axiomate 7. comprobatur, quod testatur, Animam quandoque quiescere in Deo, et idem cum eo in summo essentiae suae limite factam esse. Et per Axioma 2. in Quaest. 1 mentem a Deo non esse divisam.

A. *If the circle with its divisions by the regular polygons is (as you say) reflected in the soul from the very beginning, then the soul is divisible and multiplicable.*

Si in Anima reluceat circulus cum suis divisionibus per regularia plana ab origine, ut dicis, tunc Anima dividitur aut multiplicatur.

B. *Now the soul is neither divisible nor multiplicable.*

At Anima nec dividitur, nec multiplicatur.

C. *Therefore . . .*

Ergo . . .

A is evident because, if the circle filled it [the soul] completely (whence it is also des-

Maior constat, quia si eam impleat circulus, unde a Pla-

ignated as circle by the Platonic philosophers, though only metaphorically speaking), and if this circle were divisible into parts by the regular polygons, it follows that the soul also would be divided by the divisions of that circle.

tonicis et circulus (quamvis metaphorica locutione) dicitur, et ille circulus dividatur in partes per regularia plana, sequitur, quod et anima per divisiones illius circuli etiam dividetur.

B is confirmed by axiom 1; furthermore it is shown clearly by axiom 2 that the Creator maintained the soul as a totality before any division, wherefore from the very beginning the circle was not reflected in it, nor did it admit of the divisions of the circle by the regular polygons. But this can be stated even more lucidly in the following argument:

Minor confirmatur per Axioma 1. Praeterea per axioma 2. liquet, quod Creator obtinuerit Animam totale quiddam ante ullam divisionem, unde a primordio nec circulus in se relucebat, nec circuli divisiones per regularia plana admittebat. Sed et hoc luculentius Argumento isto sequenti declaratur:

A. *The human soul is (even in your assertion) an image of God.*

B. *Now God can neither be divided nor composed.*

C. *Therefore neither can the human soul.*

Anima humana (etiam te adstipulante) est imago Dei,

At Deus nec dividi nec componi potest,

Ergo nec Anima humana.

Replicatio, p. 34:

. . . You maintain, then, that the human soul is a part of nature and that in the soul the circle is reflected with its divisions by the regular polygons because of the fact that the soul is the image of God. But I say that the soul, at least with respect to its essence, cannot be divided from nature as a part can be divided from the

. . . Tu igitur dicis, quod Anima humana sit pars naturae, et quod in anima reluceat circulus cum suis divisionibus per regularia plana, propterea, quia animus est imago Dei: at ego dico, quod Anima, quatenus habetur ad eius essentiam respectus, non possit dividi a natura, tanquam pars a toto, iuxta illud Mercur. Trismeg. Pim. 12:

whole, according to that statement of Hermes Trismegistus (*Poimandres* 12): The mind is in no wise divided from the essence of God. Rather it is bound up with Him as is the light with the body of the sun. For we see that the solar rays are bound up with the body of the sun and cannot by any means really be divided from it, because the essence of light is a unity and cannot be divided into parts; but naturally with respect to us who abide in multiplicity we say that the soul of one man differs from that of another in number and kind, although in truth[9] all souls have a continuous relation to the *one* world-soul or the Metathron, as has the sunlight to the sun. Consequently, multiplicity really lies in matter and not in form which is nothing but a continuous emanation from God, or the Word of God, imparting life and being to all creatures. When it is withdrawn [revoked] life is destroyed, as it says in Psalm 104 . . .

Mens ab essentia Dei nequaquam divisa, sed illi potius eo modo connexa, quo Solis corpori Lumen. Videmus enim radios solares cum corpore solari esse coniunctos, et minima revera dividendos, quoniam Lucis essentia est unica, et in partes non dividenda: at vero respectu nostrum, qui in multitudine sumus versati, dicimus Animam huius ab illa alterius numero et specie differre, cum nihilominus omnes animae ad unam mundi animam seu Metathron habeant relationem continuam, ut lux Solis ad Solem. Est ergo revera pluralitas in materia et non in forma, quae nihil aliud est quam continua a Deo s. Verbi emanatio, vitam et essentiam omnibus creaturis impartiens, cuius quidem revocatione tollitur vita, iuxta illud Psalm. 104.

Replicatio, p. 35:

I therefore conclude that, as God's essence is indivisible, so also nature itself, which is His emanation into the world, is in

Concludo igitur, ut Dei essentia est indivisibilis, sic etiam ipsa Natura, quae est eiusdem emanatio in mundum, est omni-

[9] *nihilominus:* nevertheless, notwithstanding.

every respect *one single form* and indivisible in itself. And [only] in so far as God—and, hence, the functions and qualities produced in order to perfect the world—is divisible into three Persons, [only] in so far does one say that the soul, too, can be divided into various parts, whence it is sometimes the senses, now memory, now imagination, then reason, intellect, mind, and so on.

Those, then, who seek to consider the soul as it inheres in perishable things will observe with their physical eyes that it can be distinguished from the body and its properties. But he who, turning back into himself and to his centre and neglecting the external world as a deceitful shadow, penetrates into his inner gateways, he will perceive with his spiritual eyes that there is neither divisibility nor quantity in the soul and that neither numbers nor geometrical figures can be discovered in God (Who is above quantity and quality, Who has a continuous essence of soul).

. . . .

modo *unica forma* et in se indivisibilis, et quatenus Deus est divisibilis in Personas tres, inde arguendo officia et proprietates ad huius mundi perfectionem productae, sic etiam et anima in partes varias dividi dicitur, unde quandoque est sensus, nunc memoria, aliquando imaginatio, deinde ratio, intellectus, Mens etc. Qui igitur Animam considerare gestiunt rebus caducis inditam, oculis corporeis eam cum corpore eiusque proprietatibus distingui animadvertent. At qui in se et ad centrum suum revertendo, externo, quasi umbra praestigiosa, neglecto, ad interiores suos aditus penetrabit, is quidem oculis spiritualibus percipiet nec divisionem nec quantitatem inesse animae, nec in Deo (qui est supra quantum et quale, cui animae essentia continua est) numeros aut figuras geometricas posse investigari.

. . . .

But the world-soul is not on this account a circle, neither is there a circle within it; but rather by its own circular motion it encompasses and contains the universe as in the

Nec tamen propterea anima mundi est circulus, neque circulus ei inest, sed ipsa potius suo motu circulari quasi per figuram capacissimam mundum terminat atque continet, ut et

most capacious figure, and also divides it from the darkness of matter. The circle and its imaginary divisions exist, therefore, in the created passive spirit and not in the creating soul.

ab hyles tenebris dividit. Est ergo circulus eiusque divisiones imaginariae in spiritu passivo creato et non in anima creante.

APPENDIX II

Fludd on the Quaternary

Demonstratio quaedam analytica (*Discursus analyticus*) (Frankfort on the Main, 1621), Analysis of Text XXI, p. 31.

Here the dignity of the quaternary number will be discussed and I shall defend it with might and main as far as my weak intellect allows, spurred on by the insolence of the author [Kepler]. Not only has sacred theology extolled the paramount superiority of this number above the others, for which reason I feel myself moved to regard and acknowledge it as divine; but also Nature herself, the maid of the Godhead, and the nobler mathematical sciences, that is to say, Arithmetic, Geometry, Music, and Astronomy, have demonstrated its wonderful effects. Hence, when we examine thoroughly its praise in theology, we shall perceive, first of

Hoc loco in quaestionem vocatur numeri quaternarii dignitas; quam quidem, ut pro ingenii mei tenuitate manibus pedibusque defendam, urget me Authoris importunitas. Numeri igitur hujus prae caeteris excellentiam non modo summam celebravit divinitas, quo equidem eum pro divino habere et agnoscere inducor; verum etiam ipsa Natura, Divinitatis ancilla, scientiaeque Mathematicae nobiliores, videlicet Arithmetica, Geometria, Musica atque Astronomia ejus mirabiles declaraverunt effectus. Proinde, si ejus in divinitate laudes diligenter fuerimus scrutati, percipiemus primo loco, quod numerus hic quadratus Deo patri adaptetur, in quo totius Trinitatis sacro-

all, that this quadratic number is likened to God the Father in whom the mystery of the whole sacred Trinity is embraced. For the first and simple proportion of the quaternary, that of 1 : 1, denotes the symbol of the monad, the super-substantial essence of the Father, proceeding from which the second monad engendered the Son like unto Itself, and this second progression is also simple, as 1 from 1. The proportion of 2 : 2, which is the second progression from the simple numbers, denotes the Holy Ghost, proceeding from two, namely from Father and Son. These progressions in the quaternary are lucidly expressed by the ineffable name יהוה [Jahweh]: where the double He or ה [h] signifies the progression from Jod the Father and from Vau the Son, wherefore this name alone expresses the essence of God and no other is known as *Tetragrammaton*. And this is the reason why this number is called by the wise the *Origin and Source of the whole Godhead.* Nature herself, deriving her origin from the Godhead, also lays claim to this number as to her fundamental principle. And this is the same thing which the Pythagoreans proclaimed who called this number *the eternal fountainhead of nature,* as ap-

sanctae mysterium inducitur. Nam quaternarii proportio simpla et prima, unius videlicet ad unum, supersubstantiales essentiae paternae Monadis symbolum denotat, ex qua Monas secunda procedens filium genuit sibi aequalem. Atque haec processio secunda etiam simplex est ut 1 ab 1: Proportio autem duorum ad duo, quae est secunda a simplicibus processio, denotat Spiritum Sanctum, a duobus, videlicet a Patre et Filio. Quas quidem processiones in numero quaternario exprimit luculenter nomen illud inneffabile יהוה: Ubi duplex He seu ה denotat processionem a Jod Patre, et Vau Filio: Unde nomen hoc solum essentiam Dei exprimens, et non aliud dicitur *Tetragrammaton.* Atque hinc est, quod hic numerus *Caput* et *fons* a Sapientibus dicitur *totius Divinitatis.* Ipsa natura, quae ortum suum ducit a divinitate, hunc etiam numerum pro suo principio sibi vindicat: atque hoc idem est, quod cecinere Pythagorei, qui hunc numerum *perpetuum Naturae fontem* vo-

pears from the following verses which the Pythagoreans were accustomed to pronounce when taking an oath:

Pure in heart I swear to thee by the holy Four,

the fountainhead of eternal Nature, the procreator of the soul.

And while I am discussing this subject I shall say the following: the Pythagoreans did not consider duality as a number but as a blending of the unities. Consequently they declared its square to be the first even number, and not without reason; for, since the first unity signifies the divine form or *actus,* the second unity, however, the divine *potentia* or matter, *potentia* must needs emerge from darkness by virtue of the *actus*. Of these unities, now, the first was created, through the binding action of the threefold Unity, from the general [unspecified] substance of the world according to the nature of the holy Trinity. But because the first square was based upon the number 2, the progression of nature proceeded to that number 4 whose proportion to the number 2 is 2 : 1, and which is also the square of the number 2. And in this way the general, watery substance of the world was divided into four elements distinct from each

caverunt, quemadmodum ex his versibus sequentibus apparet, quibus Pythagorei iurare solebant.

Juro ego per sanctum pura tibi mente quaternium,

Aeternae fontem Naturae animique parentem.

At vero quomodo hoc sit verbo jam dicam: dualitatem Pythagorei non numerum, sed unitatum confusionem fecerunt: Unde ejus quadratum pro primo numero pari statuerunt; nec quidem hoc sine ratione: nam cum una unitas denotet formam seu actum divinum, altera vero potentiam divinam seu materiam, necesse est, ut potentia per actum ex tenebris appareat; quarum sane unitatum prior per nexum Unitatis trinae universa mundi substantia secundum proprietatem Trinitatis sacrosanctae facta est. At vero, quoniam primum quadratum erat a numero binario, ideo progressio naturae erat in numerum istum quaternarium, qui est ad numerum binarium proportio dupla, seu binarii quadratum; atque hac via divisa est universalis mundi substantia aquea in 4. Elementa ab invicem distincta. A quo quidem numero pari primario progressus sit ad Cubum primum in rerum natura, qui est numerus Octonarius, seu quaternarius duplicatus compo-

other. From this number [4] there is in the order of things a progression equal to the first, namely, to the first cube which is the number 8 or 2 × 4; this denotes the compositions of the elements just as the elements themselves, like the square, had proceeded from the number 2 which denotes simple matter and simple form as distinct from each other. From this, then, there originated the four degrees of nature which are related to the four elements, namely, being, life, sensory perception, and intelligence; the four cardinal points of the universe; the four triads in the firmament [that is to say, the four groups of three zodiacal signs, each corresponding to one of the seasons]; the four primary qualities beneath the firmament; and the four seasons. Indeed all nature can be comprehended in terms of four concepts: substance, quality, quantity, and motion. In fine, a quadruple order constantly pervades the entire nature, namely, seminal force, natural growth, maturing form, and the compost. By this we can clearly demonstrate that this number 4 should rather be chosen to distinguish and divide the humid [primal] matter than the number 3 or the number 5. Arithmetic also demonstrates the

sitiones denotans ex Elementis, sicuti ipsa elementa, tanquam quadratum, processerunt a numero binario materiam simplicem et formam simplicem ab invicem divisas denotante. Hinc igitur procedebant quatuor naturae gradus ad quatuor Elementa relati, videlicet Esse, Vivere, Sentire, Intelligere, 4 mundi Cardines, 4 in coelo Triplicitates, 4 sub coelo qualitates primae, 4 anni tempora: Imo vero tota Natura 4 terminis comprehenditur, videlicet substantia, qualitate, quantitate, et motu: quadruplex denique dispositio naturam universam implere solet, videlicet virtus seminaria, naturalis pullulatio, adolescens forma et compositum: Sed et infinitis aliis rationibus hanc Naturae quaternariae proprietatem in eandem causam respicere licet. Quibus equidem luculenter probare possumus, eam magis ad naturae [should read *materiae*] humidae divisionem et distinctionem eligendam esse, quam aut ternarium aut quinarium. Arithmetica etiam huius nu-

superiority of this number to all others. For, this science well explains, not only its twofold proportion (first, that of 1 : 2, second, that of 2 : 4) but also the origin of this proportion in that it is produced by and born from a twofold progression and proportion, namely, from 1 to 1 and from 2 to 2. The number 4 thus begins with the unity and ends in the quaternity. And indeed in this number all others are contained, for $1 + 1 = 2$, and $1 + 2 = 3$, and $3 + 1 = 4$. Thus, then, are established 1, 2, 3, and 4, in which all the mysteries of the whole world and nature herself and the extent of arithmetic are contained; for, by 3 and 4 is produced the number 7 which, *formaliter* [symbolically] considered, is downright mystical and full of secrets. From the addition of 2 and 3 there results the number 5; from $1 + 2 + 3$, the number 6; from $1 + 3 + 4$, the number 8; from $2 + 3 + 4$, the number 9; and, finally, from the summation of the entire natural progression $1 + 2 + 3 + 4$ there arises the number 10, beyond the designation of which there can be no more progress. From these progressions there arise all proportions of geometry and music, as 1, 2, 4, 6, 8, 10, and 1, 3, 6, 9, and 1, 4, 8. And from him who prop-

meri eminentiam prae aliis numeris arguit; quippe qua scientia tam duplex ejus proportio optime explicatur, quarum videlicet prior est unius ad duo, posterior vero duorum ad quatuor, quam ejusdem origo cum gemina processione et proportione producitur ac nascitur, videlicet unius ad unum et duorum ad duo. Ab unitate ergo incipit et in quaternitate definit. Et quidem in hoc numero omnes alii numeri comprehenduntur: Nam unum et unum faciunt 2 et ex $1 + 2$ exurgunt 3, atque ex tribus cum unitate 4 oritur: Sic ergo stabunt 1, 2, 3, 4, in quibus omnia totius mundi et ipsius Naturae mysteria atque Arithmetices dimensio continentur: Nam ex 3 et 4 numerus septenarius prodit, qui si formaliter consideretur plane mysticus, arcanisque plenissimus est; Ex 2 et 3 numerus quinarius additione resultat; ex 1, 2, 3 numerus senarius, ex 1, 3 et 4 numerus octonarius, ex 2, 3 et 4 numerus novenarius et tandem ex aggregatione totius progressionis naturalis 1, 2, 3, 4 numerus producitur denarius ultra cujus denominationem non est digressio. Ex his igitur progressiones omnes Geometricae et proportiones Musicae oriuntur, ut 1, 2, 4, 6, 8, 10. Et 1, 3, 6, 9. Et 1, 4, 8. Et quidem qui recte intelligit usum pro-

erly understands the use of this natural progression 1, 2, 3, 4 in formal [symbolical] speculation there will not be hidden the mystery of the seven days of creation; and why the sun was created on the fourth day; and how 3 + 4 constitute either the 7 or the 10 or the 4 among the rational numbers; and why the number 4 is the number of the day of Sabbath, viz., rest; and why the number 4 is the day of the sun; likewise, how in the true operation of nature the triad denotes and establishes the hexad and brings about six days in the work of creation. He will also be able to work out the formulae of the Critical Days and the Climactic Years. When he considers the 4 as a unity he will see, with open eyes as it were, the creation of the seven planets in the world, and many other wonders. In geometry its power is infinite inasmuch as it comprises this part of mathematics in four concepts: point, line, surface, and body. From it [the number 4] we also see emerging that aboriginal geometrical cube from the innermost part of which our author Kepler has produced all the rest just as the four elements from the womb of chaos; for, the cube, which he himself acknowledges as primordial and

gressionis istius naturalis 1, 2, 3, 4 in formali speculatione, ei non occultabitur mysterium 7 dierum creationis, et cur Sol quarto die factus sit, ac quomodo 3 ad 4 constituant numerum sive septenarium, sive denarium, sive quaternarium in numeris rationalibus, et cur numerus quaternarius sit dies Sabbathi seu requietis, et cur numerus quaternarius sit dies solis, item quomodo in vera operatione naturali tria denotent et constituant numerum senarium, diesque in creatione sex importent; dierum criticarum annorumque climacteriorum rationem explicabitur, imo vero et considerando 4 pro unitate videbitur oculis quasi apertis septem Planetarum in mundo procreationem, et multa alia mirabilia. In Geometria infinita est ejus potestas, quatenus hanc Mathesis partem 4 terminis amplectitur, nempe puncto, linea, superficie et corpore: Ab eo etiam Cubum illum Geometricum originalem procedere cernimus, ex cuius medullis ipse author noster produxit caetera, tanquam 4 elementa ex ventre Chaos: nam ex quadrati multiplicatione Cubus producitur, quem ipse agnovit esse primi-

containing the formula of everything, results from the multiplication of the square. Since this is so, I was obliged to choose in my divisions the number 4, into which the cube can be resolved as into its primary elements—namely, squares—from which, by his own admission the triangle and the pentagon are obtained. Consequently, a composite natural thing, related as it is to the cube, should be divided into quarters, viz., squares, rather than into three thirds or five fifths. For, in the act of decomposition there takes place the dissolution of the composite, viz., the cube, into four elements, that is to say, into the square; just as, conversely, in the act of generation there is a natural progression from the square to the cube. Finally the power of this number is revealed as clearly as possible in the science of music, inasmuch as it comprehends in itself the entire musical harmony. For in the double [proportion], as 1 : 2, lies the octave; in the sesquialtera, i.e., 2 : 3, the fifth; and in the sesquitertia, i.e., 3 : 4, the fourth. Furthermore, from the number 4 and its root there result all the proportions of the composite consonances [chords]. The octave, for example, stands in relation to the fifth in the

genium, et rationem totius habentem. Quod quidam cum ita sit, necesse erat, ut numerum quaternarium potius eligeremus in nostris divisonibus; quippe in quem Cubus resolvitur tanquam in prima sua elementa, videlicet quadrata, ex quibus triangulus et pentangulus secundum proprius illius confessionem eliciuntur. Proinde dividenda est potius res naturalis composita Cubo relata, in suas quartas, tanquam quadrata, quam in 3 tertias aut quinque quintas; quoniam in corruptione sit resolutio compositi seu Cubi in 4 elementa seu quadratum, sicut: a converso in generatione progressio naturalis a quadrato fit ad Cubum. Quam exactissime denique reperitur huius numeri vis in scientia Musica, quatenus ipse in se omnem Musica harmonicam comprehendit: Nam in dupla, ut 1 ad 2 consistit Diapason; in sesquialtera, ut 2 ad 3 consistit Diapente, et in sesquitertia, ut 3 et 4 Diatessaron se habet. Porro etiam ex numero quaternario et ejus radice omnes consonantiarum compositarum proportiones oriuntur; ut Diapason cum Diapente se habet in tripla, ut 2, 4, 6. Nam inter 2 et 6 proportio tripla est aggregata ex dupla, nempe 2 et 4 et sesquialtera, videlicet 4 et 6. Sed bis Diapason reperitur in quadrupla,

triple [proportion], i.e., as 2, 4, 6. For, between 2 and 6 a triple proportion is assembled from the double, namely 2 + 4, and the sesquialtera, i.e., 4 : 6. The double octave is found in the fourfold [proportion], as 2, 4, 8; the fourth, however, plus the fifth makes one octave, as 2, 3, 4. From this it can be seen that all musical proportions receive their properties from the quaternary and its root and either resolve themselves into its measures or arise from them. And, finally, if we consider mystic Astronomy we shall indeed perceive in it the whole power of the quaternary, and this most clearly; for its whole secret lies in the hieroglyphic monad which exhibits the symbols of sun, moon, the elements, and fire, that is to say, those four which are actively and passively at work in the universe in order to produce therein the perpetual changes whereby corruption and generation take place in it. The figure is as follows: [see Fig. 2]

ut 2, 4, 8. Diatessaron autem et Diapente unum constituunt Diapason, ut 2, 3, 4. Ex quibus videre licet, quod omnes proportiones in Musica ex numero quaternario, et ejus radice virtutes suas recipiant, et in ejus dimensiones vel cadant vel exurgant. Ad Astronomiam denique mysticam si respiciamus, totam equidem numeri quaternarii vim in ea perspiciamus, idque luculentissime; cum totum ejus arcanum in Monade hieroglyphica comprehendatur, Lunae, Solis, elementorum, et ignis symbola prae se ferente, tanquam quatuor illa, quae in mundo agunt et patiuntur ad inducendas assiduas in eo mutationes, quibus tam corruptiones, quam generationes in eo fiunt Figura est hujusmodi: [see Fig. 2]

In this symbolic image we see, first of all, an indication of the quaternary in the cross, four lines being arranged so as to meet in a common point. Joined with the number 3, which denotes the moon, the sun, and fire, this [quaternary] will produce the number 7,

In quo quidem symbolismo videmus primum in cruce numeri quaternarii indicium per dispositionem quatuor linearum in communi puncto, qui juncto numero ternario, Lunam, Solem et Ignem denotante numerum producet septenarium; id quod

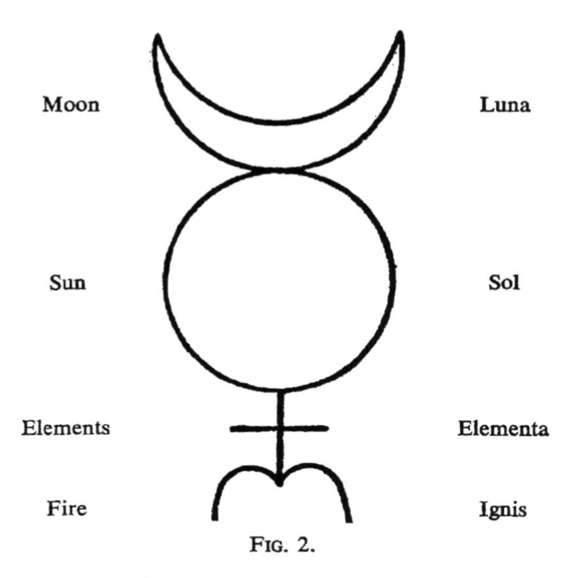

FIG. 2.

which can also be demonstrated by the four elements. And yet this number 7 is in itself none other than the quaternary considered formally.

Furthermore, even the practitioners of ordinary astronomy have esteemed this matter as of great moment: in establishing the Zodiac, they divided it into four triads. We conclude, therefore, that the wise men called this number *Tetraktys* and gave it precedence above all other numbers because, as has been said, it is the foundation and root of all other numbers. Hence all fun-

etiam ex 4 elementis praestari potest: Et tamen hic numerus septenarius in se est nihil aliud quam numerus quaternarius formaliter consideratus.

Porro etiam rem observaverunt Astronomiae vulgaris petitores magni momenti, creando Zodiacum in triplicitates 4 diviserunt. Concludimus ergo, quod Sapientes hunc numerum Tetractin appellaverint, ipsumque omnibus aliis numeris virtute praetulerint; quippe qui est fundamentum et radix omnium aliorum numerorum, ut dictum est. Unde omnia fundamenta, tam in artificialibus, quam in

damentals, both in artificial and natural things, and even in the realm divine as well, are squares, as has been explained above. It follows, therefore, that the division of a natural thing by the number 4, which is the order of nature itself, is preferable to a division by the numbers 3 or 5, which are by nature derived from the root of the quaternary and consequently subordinated to it. Finally, in dividing the earth into four parts, the water into three, the air into two, and the fire into one, one should not understand this distribution as the author [Kepler] does, as has been expounded above but with respect to the formal proportion in those elements. For I endeavour to show that the nature of the earth, since it is the basis and, as it were, the source and cube of matter, has little or nothing of form or vivifying light in itself; it is, so to speak, the vessel or matrix of nature and the receptacle of the celestial influences, so that the light that it has belongs to it more by accident than by nature, inasmuch as it [the earth] is very far removed from the source of light and is the coldest of all elements, and this in the fourth degree; water is also cold but to a lesser degree. for this reason it [the earth]

naturalibus, imo vero et in divinis quadrata sunt, quemadmodum in superioribus declaratum est. Sequitur ergo, quod praestantior sit rei naturalis per numerum quaternarium divisio, qui est ipsius Naturae ordo, quam per ternarium aut quinarium, qui natura sunt radici quaternariae et per consequens ipsi quaternario postponendi. In divisione denique terrae in 4 partes, aquae in tres, aeris in duas, et ignis in unum, distributio illa non est intelligenda authoris more, ut supra declaratum est, sed respectu proportionis formalis in illis elementis: namque demonstrare nitor quod natura terrae, quatenus est basis et quasi fons ac cubus materiae, parum aut nihil habeat formae seu lucis vivificae in se et quod fit quasi vas seu matrix Naturae, atque influentiarum coelestium receptaculum, ita est illa lux, quam habet, magis ei adsit per accidens, quam a natura, quatenus longius distat a fonte lucido, et est omnium elementorum frigidissimum, idque in gradu quarto; sicuti aqua etiam frigida est, sed in gradu remissiori; quare unicum lucis

admits of only *one* degree of light into itself and so also in the case of the others. The wise ought therefore to understand rightly before condemning rashly.

gradum in se admittit; et sic in caeteris. Sapientis igitur esset recte intelligere, priusquam inconsulte condemnare.

APPENDIX III

The Platonic and Hermetic Trends: Johannes Scotus Eriugena (810?–?877)

The controversy between Kepler and Fludd is connected, from the point of view of the history of ideas, with the existence in the Middle Ages of two different philosophical trends which I may designate briefly as the Platonic and the alchemistic (or hermetic). Between these two trends there are, on the one hand, important points of agreement and even intermediary or transitional stages; but, on the other hand, there existed between them fundamental differences that seem to me to be more than mere shades of opinion. For the Platonist, the life of the Deity which he conceives in a more or less pantheistic spirit, that is to say as identical with the totality of the world, consists of a cosmic cycle which begins with the emanation from the Godhead first of the "ideas" and "souls," then of the corporeal world, and ends with the return of all things to God. The idea of the *opus* and its result, and thus the idea of transmutation, is foreign to the Platonist. The final stage of the cycle is identical with the initial stage,[1] and this process continues for ever and ever. What, then, is the meaning of this eternal cycle if it does not lead to any result? To this question the Platonist gives the answer: beauty. The prime cause of the cycle is unchangeable and unmovable,

[1] Scotus Eriugena: "Finis enim totius motus est principium sui; non enim alio fine terminatur nisi suo principio a quo incipit moveri."

drawing things back into itself solely by virtue of its beauty.[2] The cycle pursues a self-sufficient beauty guaranteed by "rules of the game" which are determined once and for all, and it needs no result. The soul of the individual can do nothing but fit itself into this cosmic cycle in order to become a participant in the beauty of the universe.[3] This is the purpose of contemplation which always begins with melancholia, with the homesickness of the soul for its divine origin. (The parallel to the "melancholia" of the Platonists is the alchemists' *nigredo.*)

Despite all my respect for the philosophy of the Platonists, it seems to me that the attitude of the alchemists, with their *filius philosophorum* as a symbol of transformed totality, is closer to modern feeling. In particular, the Platonic idea of a primal cause that produces effects but cannot be affected in turn is not acceptable to the modern scientist, who is accustomed to the relativity of reciprocal effects. I believe also that this idea can hardly stand the test of psychological analysis; it seems to be determined by the particular and by no means generally valid psychology of its authors, a psychology that showed a tendency to deny the reciprocity between ego-consciousness and the unconscious.

The Platonists, as we have seen in Kepler's case also, favored in general a trinitarian attitude in which the soul occupies an intermediary position between mind and body. It may be of considerable interest to know, however, that in the

[2] Scotus Eriugena: "Ita rerum omnium causa omnia, quae ex se sunt, ad se ipsum reducit, sine ullo sui motu, sed sola suae pulchritudinis virtute."

[3] In the Renaissance Platonism of Leone Ebreo and Marsilio Ficino the cycle appears specifically as the *circulus amorosus.* According to these authors the bliss of love lies in the fact that the lovers insert themselves into the cyclical current pervading the cosmos. The conception of love is broad enough to include both the desire for knowledge as *amor intellectualis dei* and the ecstatic states of the religious prophets as *amor coelestis.* For the alchemical parallels to this *circulus amorosus,* cf. the series of pictures in Jung's "Psychology of the Transference" (in *The Practice of Psychotherapy,* New York and London, 1954); and fig. 131 in his *Psychology and Alchemy* (New York and London, 1953), which corresponds to the beginning of this circle.

1. *Natura creans nec creata. Origo:* God the Father

2. *Natura creans creata.* "Ideas": God the Son

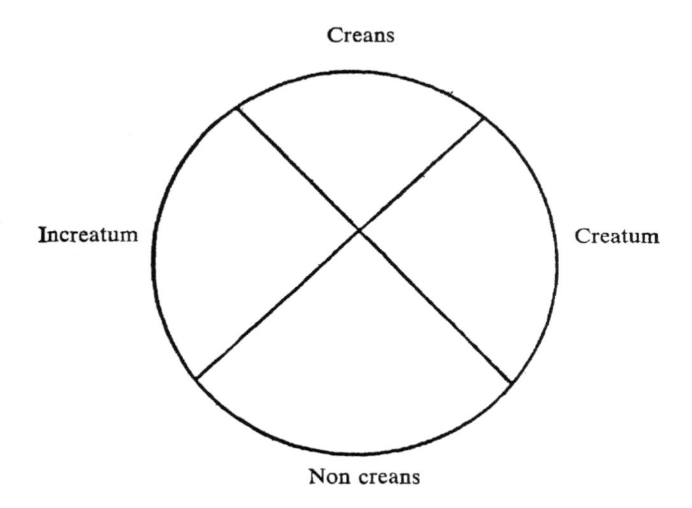

4. *Natura nec creata nec creans.* Goal: *theosis (deificatio)*

3. *Natura creata nec creans.* "World": products of emanation, the corporeal world, matter; *theophaniai,* the Holy Ghost. "God has created Himself in the world"

FIG. 3. Quaternity as conceived by Scotus Eriugena in *De divisione naturae* earliest Platonic thinker of the Middle Ages, Scotus Eriugena, the idea of quaternity can also be found. In his work *De divisione naturae* (862–66), he introduces two pairs of opposites: a pair of active principles, viz., the *creans* (that which creates) as opposed to the *non creans* (that which does not create); and a pair of passive principles, viz., the *creatum* (that which is created) and the *non creatum* (that which is not created). By the aid of this terminology, which is very attractive to the mathematically minded, Scotus arrives at his four natures, a conception that may be illustrated by the schematic drawing in Fig. 3, which also reveals the connection of Eriugena's system with the Platonic cycle of emanation and re-absorption. In identifying Stages 1 to 3 of the cycle with the three Divine Persons, Scotus Eriugena attempted to compromise with the dogma of the Church. In the case of the fourth stage, however, that of the *natura nec creata nec creans,* he seems to have found himself in an embarrassing position. As a Platonist he could not do as the Hermetic philosophers did and allow a transformation of the whole to appear simultaneously with this fourth stage. Since he wanted to return to the point of departure where no fourth Divine Person was at his disposal, he could think of nothing better than to act as though the *natura nec creata nec creans* were the same thing as the *natura creans nec creata* at the beginning, for which assumption no satisfactory reason is given.[4] To the question of what has happened to the fourth Person, therefore, the answer must be in the particular case of Scotus Eriugena: "He has disappeared in an identification with the first."

[4] It was Professor Markus Fierz who called my attention to this point.

INDEX

INDEX

Made in the USA
Columbia, SC
31 October 2019

82478448R00154